When Tradition Turns into Innovation

When Tradition Turns into Innovation

How firms can create and appropriate value through tradition

ANTONIO MESSENI PETRUZZELLI AND VITO ALBINO

CP
CHANDOS PUBLISHING

Oxford Cambridge New Delhi

Chandos Publishing
Hexagon House
Avenue 4
Station Lane
Witney
Oxford OX28 4BN
UK
Tel: +44 (0) 1993 848726
Email: info@chandospublishing.com
www.chandospublishing.com

Chandos Publishing is an imprint of Woodhead Publishing Limited

Woodhead Publishing Limited
80 High Street
Sawston
Cambridge CB22 3HJ
UK
Tel: +44 (0) 1223 499140
www.woodheadpublishing.com

First published in 2012

ISBN: 978-1-84334-664-7 (print) (Chandos Publishing)
ISBN: 978-0-85709-173-4 (print) (Woodhead Publishing)
ISBN: 978-0-85709-382-0 (online)

British Library Cataloguing-in-Publication Data
A catalogue record for this book is available from the British Library.

Typeset by RefineCatch Ltd, Bungay, Suffolk
Printed in the UK and USA.

Printed in the UK by 4edge Limited - www.4edge.co.uk

Contents

List of figures and tables

Figures

Tables

Acknowledgements

The writing of this book has been a journey that has allowed us to investigate a fascinating issue – the relationship between tradition and innovation – and to increase our awareness of how the past shapes the future, in life as in business. We would like to express our gratitude to Tommaso Savino, who has provided a significant contribution to the collection of information and data about the analysed cases, as well as helping to give a better understanding of what tradition means.

About the authors

Antonio Messeni Petruzzelli is a lecturer in Innovation Management at the Politecnico di Bari, where he teaches technology strategy, and is co-founder of the Innovation Management Group. He gained his laureate degree in Business Engineering at the same university. After a two-year period as organisational analyst at Eni Group, he was awarded a PhD in Innovation Management at the Politecnico di Bari with a dissertation entitled 'Knowledge Relationships and Proximity: The analysis of knowledge gatekeepers'. In 2008, he was visiting scholar at the IESE Business School, in the Department of Strategic Management, where he concentrated on the role that relational attributes play in the relevance of university–industry collaborations. In 2011, he collaborated with the Technical Committee for the Development of the Strategic Plan of the Mechatronic Technological District of the Apulia region (MEDIS).

His main research interests concern the area of innovation management, including themes such as knowledge search and recombination, inter-organisational relationships and system dynamics modelling. Within these topics, he has published several articles in international journals (e.g. *Technovation*, *Scandinavian Journal of Management*, *Journal of Knowledge Management* and *International Journal of Production Economics*) and presented papers at international conferences. Currently, he is a contributing writer to *Wisepreneur*, an on-line magazine relating to innovation, leadership and entrepreneurship and a member of several international associations, such as the Academy of Management Association, the Strategic Management Society and the DRUID Society.

The author may be contacted at:
Politecnico di Bari
Dipartimento di Ingegneria Meccanica e Gestionale
Viale Japigia 182
70126 – Bari – Italy
E-mail: *a.messeni.petruzzelli@poliba.it*

Vito Albino is Professor in Innovation and Project Management at the Politecnico di Bari in Italy. At the same university, he is the Founder and Director of the Knowledge Management Lab, co-founder of the Innovation Management Group and, at present, in charge of the industrial liaison office. Vito is also Vice-President of the Board of Consiglio di Territorio Sud Continentale, a group of experts set up from the UniCredit S.p.A., the largest Italian Bank, to sustain the development of the business sector in southern Italy, and is a member of the board of the Mechatronic Technological District of Apulia region (MEDIS).

After gaining his laureate degree in Mechanical Engineering, Vito became Visiting Researcher at the University of Cincinnati from 1986 to 1987 before beginning his academic career as a researcher at the University of Basilicata in 1988. He became Associate Professor there and then moved to the Politecnico di Bari as Professor. While Visiting Professor at the University of South Florida (Tampa), he gave lectures and seminars in many universities in the USA and in Europe.

In 2003, he co-founded the European Sustainability Group, a group of organisations, researchers and consultants in Europe who are actively working to create a sustainable world, through creatively learning together, sharing insights and developing practical capacity for sustainability in the organizations they work with. Commencing in the same year, he also chaired the Dipartimento di Ingegneria Meccanica e Gestionale at the Politecnico di Bari for six years, and from 2007 to 2009 he was the President of the Board of the Italian Association of Business Engineering (AiIG). Also in 2003, Vito served as a member of the Scientific Committee of the Italian Ministry of University and Research for the internationalisation of the Italian university system, and two years later as an expert of the European Union in the area Innovative Environment for the project Information Society Network (URBACT Programme). In 2008 he was the Italian scientific coordinator of the research project 'Sustainability Knowledge Integration Mechanisms (SKIMs) and Green Product Strategies in Canada and Italy: A Cross-Cultural Perspective' with a grant from the Social Sciences and Humanities Research Council of Canada. In 2010 he became co-director of the course 'Sustainable Strategies and Operations' at Columbia University and had scientific responsibility for the project 'A review and synthesis of research on the use of alternative energy sources in cement manufacturing', with a grant from the Network for Business Sustainability (Canada).

His research fields are related to innovation processes in production chains and industrial clusters, with particular interest in environmental aspects.

The author may be contacted at:
Politecnico di Bari
Dipartimento di Ingegneria Meccanica e Gestionale
Viale Japigia 182
70126 – Bari – Italy
E-mail: *v.albino@poliba.it*

Preface

Innovation, generally, has been conceived as the most important driver leading to firms' competitiveness and economic growth. Starting from this widely-accepted assumption, this book aims to provide a new perspective on the innovation process by focusing on the important role that may be played by old knowledge and capabilities. Specifically, we propose that general managers and corporate executives should be able to act as the Roman god Janus, who had two sets of eyes, one pair focusing on what lay behind, the other on what lay ahead. As Janus, managers must constantly look backward, attending to the products and processes of the past, while also gazing forward, preparing for the innovations that will define the future. This balancing act requires executives to explore new opportunities, as well as to exploit existing capabilities that have contributed to their firms' success and competitiveness. In line with the above arguments, we discuss the strategic relevance of tradition as a source upon which companies may shape competition by creating new value for customers and appropriating it. Tradition is defined as the whole set of competencies, knowledge, values and culture that characterise a specific firm, as well as a territory and an age. Thus, relying on tradition, companies may innovate by introducing new products and processes and differ from competitors in terms of both meaning and content.

This book may be considered as a useful support for both managers and executives constantly dealing with the necessity of differentiating business models by finding new innovative solutions for providing uniqueness to their products and processes. Moreover, the proposed attention toward the rediscovery of the past contributes to shed new light on the value of searching knowledge over time, thus calling for further research on the relationships between old and new.

The first chapter of the book proposes a review and criticism of the actual innovation approaches, highlighting how the actual economic scenario imposes the necessity of rethinking innovation. In Chapter 2, the concept of tradition is analysed by presenting its meaning and origin, as well as why it may assume the semblance of a competitive driver. In Chapter 3, we discuss the strategic role of tradition, explaining how it may be embedded both in products and processes, and focusing on two distinct drivers – technology and design. Furthermore, the strategic role of tradition is investigated from the value creation and appropriation perspectives, showing its relationship with firms' performance. This is followed in Chapter 4 with the presentation and discussion of several cases of companies strategically exploiting tradition. Firms are located worldwide and operate in different industries, such as pharmaceutical, food, beverage and financial services. Finally, in Chapter 5 the main ideas, concepts and approaches emerging from the book are summarised, thus tracing the pattern of an alternative innovation strategy.

1

Re-thinking the innovation approach

Abstract: This chapter presents a review and criticism of the actual innovation approaches, highlighting how the social and economic scenario imposes the necessity of rethinking innovation and consumption models. Specifically, we discuss how the recent crises – which together affect finance, food and climate change and their implications for human development – are forcing organisations to find new solutions and models for responding to emerging needs and expectations. In this regard, we elaborate on the important role that may be played by traditional knowledge as a source of inspiration for innovation, since creativity can find a reliable support in what society has found to be suitable in the past for its development needs.

Key words: triple crisis, innovation, tradition.

For most members of the Organisation for Economic Co-operation and Development (OECD), the financial crisis that started in 2007 triggered a global economic downturn, which led to the worst fall in economic activity (GDP) since the Great Depression. At first this resulted in falling economic growth rates followed by a real economic decline in many countries. In Table 1.1, the GDP for some regional areas and for the years 2004–2009 is reported.

Table 1.1 **Gross domestic product* (expenditure approach)**

Year	Euro area (16 countries)	European Union (27 countries)	United States	OECD
2004	9,023	12,601	11,812	32,904
2005	9,489	13,218	12,580	34,844
2006	10,292	14,360	13,336	37,397
2007	10,937	15,263	13,995	39,526
2008	11,405	16,006	14,297	40,846
2009	11,159	15,676	14,044	40,051

Source: OECD, *http://stats.oecd.org/Index.aspx?datasetcode=SNA_TABLE1 2011*

* $US billion, current prices, current purchasing power parities (PPPs), OECD base year

As shown in Table 1.2, the unemployment rate in the European Union (EU27) and in the United States has strongly increased in the past two years, achieving a double-digit value.

As a consequence of this economic downturn, the governments of developed countries generally adopted intervention policies resulting in an increase in public debt as shown in Table 1.3.

Table 1.2 **Unemployment rate (per cent) in some regional areas**

Year	European Union	United States	OECD
2005	8.9	5.1	6.6
2006	8.4	4.7	6.1
2007	7.3	4.4	5.6
2008	6.7	5.1	5.9
2009	8.3	8.5	n.a.
2010	9.6	9.7	n.a.

Source: Eurostat and OECD

Table 1.3 **Public debt in some regional areas***

Year	Euro area	United States	OECD
2008	75.8	70.4	79.0
2009	89.3	83.0	90.3

Source: Global Finance, http://www.gfmag.com/tools/global-database/economic-data/10394-public-debt-by-country.html 2011

*Public debt as per cent of GDP in some regions.

The economic crisis that has recently emerged, and some of its consequences, will last for the foreseeable future; and it will be a decade in which our civilisation will be fully aware of a more dangerous and global crisis: climate change. Global climate change started a long time ago with the industrial revolution, which changed the way in which goods are produced and are available for a wider set of consumers. The faster rate of transformation of natural resources into products progressively caused an increased exploitation of such resources and the emission of more pollutants (i.e. greenhouse gases) into the environment. The earlier industrial development and the population growth of western countries, and more recently of countries such as Brazil, China, India and Indonesia, is now impacting on the environment and causing climate change in different geographical areas of the world. In particular, some poor countries such as Congo and Somalia,[1] are experiencing the negative impact of this change, with huge damage and a wider problem of crops and food production.

For the past few decades, some world organisations (United Nations, World Bank, etc.) have been dealing with ways to control such a change. Some measures, based on a small number of indicators, can support the analysis of the crisis. These indicators are evaluated with reference to

OECD countries which are historically responsible for such a crisis. Carbon dioxide (CO_2) makes up the largest share of greenhouse gases, considered as mainly responsible for the change in weather conditions in many countries. In Table 1.4, emissions of CO_2 caused by burning oil, coal and gas for energy use are reported with reference to the OECD countries. Carbon dioxide also enters the atmosphere from burning wood and waste materials and from some industrial processes such as cement production. However, emissions of CO_2 from these other sources are a relatively small part of global emissions and are not included in the statistics shown in Table 1.4. The amount of electricity generated by a country reflects the natural resources, imported energy, national policies on security of energy supply, population size and electrification rate, as well as the stage of development and rate of growth of the economy in each country. In Table 1.4, data on electricity generation from fossil fuels, nuclear, hydro (excluding pumped storage), geothermal, solar, biomass, etc. are given with reference to OECD countries. Another indicator of the environmental crisis is related to municipal waste generation. The amount of municipal waste generated in a country is related to the rate of urbanisation, the types and patterns of consumption, household revenue and lifestyles. While municipal waste is only one part of the total waste generated in each country, its management and treatment often absorbs more than one third of the public sector's financial efforts to abate and control pollution. The main concerns raised by municipal waste are the potential impact of inappropriate waste management on human health and the environment (soil and water contamination, air quality, land use and landscape). In Table 1.4 the amount of municipal waste generated in OECD countries is reported.

The environmental crisis is forcing governments to change production and consumption models while trying to look for

Table 1.4 **Some indicators of the environmental crisis for OECD countries**

Indicators	1990	2000	2007
CO_2 emissions from fuel combustion (million tonnes)	11.073	12.492	13.001
Municipal waste generated (million tonnes)	489	590	623
Electricity generated (Terawatt hours)	7.568	9.618	10.645

a new and more appropriate balance with the natural environment (exploitation of resources). This is not an easy balance, since our societies are experiencing contemporary critical situations. In fact, the concept of triple crisis[2] has been proposed (Tarp and Addison, 2010) to jointly take into account the finance, food and climate change problems, and their implications for poverty, inequality and human development.[3] Often the public debate about the economic and environmental crises recognises a common cause in the poor ethics with which, in developed economies in particular, a higher welfare is expected with a faster pace.

In general, the progress of science and technology is considered as a driver of civilisation and has important effects on the welfare of mankind. However, the ethical behaviour of individuals and of institutions/governments in quickly exploiting available resources (natural and man-made, such as knowledge) is often responsible for undermining the progress of civilisation. There are few arguments in support of such an idea. This is mainly mirrored in the increasing public deficit in Western countries, hence transferring the problem to the next generations and calling for a quick reaction. Unfortunately, GDP in Western countries is not increasing as fast as governments would wish in order to mitigate the negative impact of huge public debt. A high level of unemployment is another sad effect of such a situation.

The appropriate and ethical use of all resources is therefore a general and primary question for mankind; and this is still true even if the economic context turns out to be positive. The present scenario could represent an opportunity to foster a shift toward a new society where more suitable production and consumption models can give the right value to natural and man-made resources. Today, more than ever, in a context of climate change and financial crisis, it has become clear that our global community has to adopt more sustainable lifestyles to both reduce the use of natural resources and CO_2 emissions and move toward low-carbon societies and a green economy (UNEP, 2011). This is crucial in order to decouple economic growth from environmental exploitation and degradation, in both developing and developed countries, as well as to create the 'space' for the poor to meet their basic needs. In the past few years, the debate about the concept of a new and green economy has dramatically grown.

Several factors force the political and academic decision makers to consider new economic principles and values. Some of these factors have emerged since the Rio Conference, which was held in 1992 in Rio de Janeiro and is known as the Earth Summit II. During the Opening Session, Maurice Strong, the Secretary General of the Conference, gave the introductory talk stating that industrialised countries have: 'developed and benefited from the unsustainable patterns of production and consumption which have produced our present dilemma. It is clear that current lifestyles and consumption patterns of the affluent middle class, involving high meat intake, consumption of large amounts of frozen and convenience foods, use of fossil fuels, appliances, home and workplace air-conditioning, and suburban housing are no more sustainable. A shift is necessary toward lifestyles less geared to environmentally damaging consumption patterns.' As an immediate conclusion, the economic model pioneered by today's industrial countries is not viable for the world as a whole. In fact, today's economy is profoundly out of balance with the world's natural system. We realise that solving environmental problems will entail substantial costs for some of today's industries. However, it will create thousands of new companies and millions of jobs, laying the foundation for a green economic transformation.

As stated in Chapter 4.3 of Agenda 21,[5] 'the major causes of the continued deterioration of the global environment are the unsustainable patterns of consumption and production, particularly in industrialised countries, which is a matter of grave concern, aggravating poverty and imbalances.' The Marrakech Process[6] has been proposed as the global multi-stakeholder process charged with promoting sustainable consumption and production (SCP) and to work toward a Global Framework for Action on SCP – the so-called 10-Year Framework of Programmes on SCP (10YFP). The major goals are to assist countries in their efforts to 'green' their economies, to help corporations to develop greener business models and to encourage consumers to adopt more sustainable lifestyles. The United Nations Environment Programme (UNEP) and the United Nations Department for Economic and Social Affairs (UN-DESA) are the leading agencies of this global process, with the active participation of national governments, development agencies, private sector, civil society and other stakeholders (see *http://esa.un.org/marrakechprocess/*).

Sustainable consumption and production (SCP) can be defined as the use of services and related products, which respond to basic needs and bring a better quality of life while minimising the use of natural resources and toxic materials as well as the emissions of waste and pollutants over the life cycle of the service or product so as not to jeopardise the needs of future generations (OECD, 1997). Innovation processes aiming to support SCP models also encourage resource and energy efficiency, sustainable infrastructure, green jobs and a better quality of life. Innovation can accelerate the achievement of long-term sustainable development by reducing future economic, environmental and social costs, strengthening economic competitiveness and reducing poverty. However, at the world level, UNEP – whose aim is to assist governments in 'greening' their economies by reshaping and refocusing policies, investments and spending toward a range of sectors[7] – supported the Global Green New Deal (GGND). Since the financial and economic crisis has triggered an unprecedented policy response, it should make a major contribution to reviving the world economy, saving and creating jobs and protecting vulnerable groups. It should promote sustainable and inclusive growth and the achievement of the Millennium Development Goals, especially ending extreme poverty by 2015. Also, it must reduce carbon dependency and ecosystem degradation; these are key risks along a path to a sustainable world economy.

UNEP has considered the G20 a powerful organisation, better able to coordinate the economic recovery at the world level. The notion that investments in environmentally-friendly technologies and infrastructure will yield significant and more sustainable economic growth opportunities has come to be known as the Green Stimulus Package. This investment stimulus should be able to impact on production and consumption models through a series of innovations in the entire economy of all acting countries. Driven by innovation, trends in production and consumption models are already emerging. In particular, some trends are consistent with the actual dynamics of the globalisation phenomena.

In fact, which kind of globalisation could be suitable seems to be central to the current debate (e.g. Guptara, 2010). Related to this issue, the evergreen problem of how to balance global and local visions at the level of country or company is going to be explored (Kanter, 1995). As a specific but relevant issue, in the past decade globalisation transformed companies' supply chains and now it is common to have products whose components have been manufactured and assembled on different continents. However, this approach is now under careful evaluation since environmental and operational problems have shown some weaknesses which ask for radically reinvented supply chains (see for instance Lee, 2010; de Treville and Trigeorgis, 2010).

Similarly, but at the social and cultural level, there is a profound contradiction that many people now live with thanks to globalisation. Individuals think that they belong to a particular national or regional culture and believe that they must cherish it. At the same time, they also consume from the global cultural supermarket (Mathews, 2000). They eat cornflakes for breakfast or sushi for dinner and buy, play, and even be anything in the world (objects, music, religion, respectively). Since identity formation is the process of the development of the distinct personality of an individual, regarded as a persisting entity (known as personal continuity), in a particular stage of life in which individual characteristics are possessed, each person is then recognised or known (such as the establishment of a reputation) on the basis of these characteristics (Friedman, 1992). This process defines individuals and shapes identity, including a sense of continuity, a sense of uniqueness from others and a sense of affiliation. This may be through individuation, whereby the undifferentiated individual tends to become unique, or undergoes stages through which differentiated facets of a person's life tend toward becoming a more indivisible whole.

Then, it seems reasonable to ask if people can choose aspects of their life and culture from all of the world while still maintaining their 'home'.[8] However, the above contradiction often creates tensions and stress, and individuals make efforts to recover their identity. One example of how they can do this is to look at their past. In fact, approaching traditional products and processes, as well as cultural values, could be the way to recover the past and then the identity. In such a complex scenario, new production and consumption models can emerge as a result of an evolutionary process where innovation is the main driver of economic and social change. We expect innovation to be able to destroy inefficient market and organisation structures, as well as to create a sustainable balance between what we are and what we want to be from the cultural and social point of view.

In *Capitalism, Socialism and Democracy*, Joseph Schumpeter (1942) argued that:

> . . . the fundamental impulse that sets and keeps the capitalist engine in motion comes from the new consumers' goods, the new methods of production or transportation, the new markets, the new forms of industrial organisation that capitalist enterprise creates.

Entrepreneurs discover and implement innovations that 'incessantly revolutionise . . . the economic structure from within, incessantly destroying the old one, incessantly creating a new one', a process that he labelled creative destruction. The innovation process is creative in the sense that it creates value (increases consumer welfare) and is destructive in the sense that it reduces or eliminates returns to capital and labour involved in the production of obsolete products. In particular, as technological innovation progresses at a very fast pace, it impacts upon consumer welfare and returns to capital and labour in a dramatic way. Technology seems to have an emerging role in designing society and economy. This sense of technology's power as a crucial agent of change has a prominent place in the culture of modernity. It belongs to the body of widely shared tacit knowledge that is more likely to be acquired by direct experience in the use of technology than by the transmittal of explicit knowledge (Marx and Smith, 1996). Computers, telecommunications, jet aircraft, nuclear weaponry, antibiotics, nanomaterials and biogenetic engineering are some examples of radically new, science-based technologies which have transformed our way of life.

For some three centuries, direct firsthand experience of that power has been a common and universal feature of life in developed and developing countries. The collective memory of Western culture is well stocked with lore on this theme, i.e. with the accumulated tradition, fact, or belief about technology. The role of the mechanic arts as the initiating agent of change pervades the received popular version of modern history. Chemistry, electronics, information technology and biotechnology are the next agents of change. The structure of such popular narratives embedded in Western culture demonstrates the efficacy of technology as a driving force for change: a technical innovation suddenly appears and introduces significant improvements. It is noteworthy that in the hard technology determinism these narratives direct attention to the consequences rather than the genesis of inventions. We can observe also that the emergence of new ideas in science and technology has its roots in the collective memory of people.

A historical starting point for this idea (Rosenberg and Birdzell, 1986) is the marked acceleration in the rate of technical progress that occurred in the West in the seventeenth and eighteenth centuries. Historians believe that the propensity to innovate came to the fore at that time in the British Isles, in the North American colonies, and in Western Europe for a great variety of grounded reasons. Some are firmly established on the particular efficacy of certain material, geographic, demographic and socio-economic preconditions:

- access to raw materials or markets;
- the existence of a mercantile capitalist economy;
- the presence of the profit motivation;
- the accumulation of capital;
- the availability of a teachable, exploitable labour force.

Others contend that the principal causes are associated with intellectual, cultural, or ideological factors:

- the extent of secular learning;
- the existence of a reservoir of entrepreneurial or financial skills;
- the diffusion of scientific rationalism;
- a religion-based work ethic;
- or an artisanal ethos.

Although it seems likely that the true reason is to be found in some distinctive combination of these factors, no-one can determine the recipe for the special propensity to innovate that initially developed in Western society at that time. Thus agency, as conceived by 'soft' technological determinism (Marx and Smith, 1996), is deeply embedded in the larger social structure and culture. This concept seems to be strongly confirmed by the more recent history where technological innovation operates as a major driving force of the present time. Its power to change society may be derived from certain specific socio-economic and cultural situations, in other words, from the lore of the same society. Once technology has been developed, its determinative efficacy may then become sufficient to drive the evolutionary process. Recognition that the technological structure is inextricably entwined in the activities of any society can help in the understanding of the connection between change in the technological structure and changes in the socio-economic order (Heilbroner, 1996).

In *Toward a Rational Society*, Habermas (1970) is concerned with how societies can employ previous conceptions to exert conscious control over the practice involved in technological development. This development is considered as an essentially human enterprise in which the people who create and employ technology are driven by goals and judgements about public and private goods. They act following certain culturally accepted norms and are sanctioned by politically legitimate forms of power (Bimber, 1996). The major critique pointed out by Habermas is that industrial societies have developed an overreliance on norms of efficiency and productivity in directing the conduct of this enterprise. By adopting reductionist values as references for decision-making about technology, they exclude other ethical criteria, thereby producing a self-correcting process that operates autonomously of larger political and ethical contexts. This social subsystem, made up of people pursuing the rationalisation of life through the creation of technology, is out of the control of the entire society in terms of ethical and normative judgements. The worst case occurs when society makes a decision to adopt as its own the technologists' standards of judgement. Then, an acceptable technology change has to be consistent with the values that drive the society. This is particularly true when companies are asked to become socially responsible, to include in their strategies and operations values associated with the environment and ethics. Within this frame of reference, innovation can contribute effectively to the design of new production and consumption models.

Innovation can be considered as the *cause* of knowledge accumulation, but at the same time it can be the *effect* of knowledge accumulation. In fact, innovation is driven by the accumulation of experience and scientific (both theoretical and empirical) knowledge. We may also assume (Steinmetz, 2008) that history affects market opportunities (historic dependencies), i.e. previous actions and outcomes determine the range of available actions and outcomes. This means that a firm's investment in R&D does not only increase the chance of making a discovery, it also increases the knowledge stock. This knowledge stock is a measure of a firm's past R&D effort and allows for model learning – i.e. a firm's past experiences add to its current capabilities – and organisational forgetting. Learning may occur when the innovation activities of a firm are adjusted due to past experiences, or when innovation projects are cumulative, i.e. sequential and building on each other. Organisational forgetting, on the other hand, is a phenomenon that has been shown in some studies (see for example Argote, Beckman and Epple, 1990 or Benkard, 2000). Based on the idea of knowledge capturing the experience of the workers, organisational forgetting would be the result of turnover and layoffs. The role of knowledge stock is essential in the technological paradigm underlying search activities (Mazzucato and Dosi, 2006). For instance, referring to the pharmaceutical industry, Gambardella (1995) identifies two subsequent technological paradigms. In the first, dominant before the 1980s, the search for innovation was carried out principally through 'random screening', driven by relatively tacit search heuristics and involving a great deal of serendipity (e.g. the search for one therapy leading to the unexpected discovery of a different one). A different paradigm has latterly become dominant, which rests on major theoretical advances in molecular biology and jointly in biotechnology and computational chemistry. This approach has made the search process more 'guided', and also involves in the search process an exploitation of the knowledge stock and therefore a higher degree of path dependency. Such experimental development is systematic work, drawing on existing knowledge gained from research and/or practical experience, which is directed toward producing new materials, products or devices, to installing new processes, systems and services, or to substantially improving those already produced or installed. Then, the stock of knowledge available to a firm can make the difference in terms of innovation performance. Since part of the knowledge stock is embedded in tradition, this can have a significant impact on innovation. Traditional knowledge usually refers to the long-standing traditions and practices of certain communities. It also encompasses the wisdom, knowledge and teachings of these communities and, in many cases, traditional knowledge has been orally passed from generation to generation. Some forms of traditional knowledge take the shape of stories, legends, folklore, rituals, songs and even laws. Other forms of traditional knowledge are often expressed through different means.

Traditional knowledge includes pre-existing, underlying traditional culture, or folklore and literary and artistic works created by current generations of society which are based on, or derived from, pre-existing traditional culture or folklore. Traditional culture and folklore tends to be trans-generational, old and collectively 'owned' by groups or communities. Often traditional culture and folklore is of anonymous origin and expressions of this pre-existing traditional culture are generally not protected by current intellectual property laws and are treated as being in the public domain. In contrast, some products based upon, derived from or inspired by traditional culture or folklore may incorporate new elements and contents. Hence these products may be 'new' products with a well-defined producer. Such products may include a new interpretation, arrangement, adaptation or collection of a pre-existing cultural heritage that is in the public domain. Recently, traditional knowledge has been drawn into the debate related to access to medicines, access to food, the need for poverty alleviation and related issues that affect the livelihoods and welfare of people worldwide. It has been argued that traditional knowledge is an important source of health security, food security and livelihood security for the world's poor. As such, it has been argued that the benefits of sharing and actively disseminating traditional knowledge outweigh the benefits of protection and recognition of ownership.

The open access movement in science and the humanities, in the media, arts, entertainment and communications communities, has shown that traditional knowledge can be an essential and public source of cultural creativity. Through traditional knowledge and resources, a significant contribution to the welfare of mankind has been obtained; for instance, as the basis of much of the world's food system and naturally-derived medicines. Another important example of traditional knowledge stems from what the classical arts have developed during ancient times. The idea that the classical arts could be an extraordinary source of inspiration for all kinds of art has been repeatedly confirmed during history (Settis, 2004). The Greek or Roman monuments or masterpieces have been the never-ending frame of reference and models for architects and artists, who have created whilst having in their mind the *tòpoi*, i.e. some standards used in the representation of tales, human feelings, myths and so on. For instance, one of the most exciting and extraordinary periods of artistic production, the Renaissance, owes a great debt to classicism since at that time it reinvigorated the 'classical mind'; a wonderful and astonishing interpretation being the creation of sublime opera.

A similar impact has been recognised in the field of civilisations, where tradition represents the result of experience stratification or, in other words, the persistent values which endure and are able to flourish again, even after centuries, generating new ideas and opportunities (Braudel, 1985). As this brief overview of some fields of human progress has shown, tradition can be a source of inspiration for innovation since creativity can find a reliable support in what society has found to be suitable in the past for its developmental needs. It represents the frame of reference which provides some guidelines about what has survived in the economic, social and cultural evolution encompassing therefore the idea of sustainability. However, this doesn't mean it has to last in an unchanged way. On the contrary, tradition is an asset that could be continuously explored and expanded.

Notes

1. See FAO (2010) *The State of Food Insecurity in the World 2010*, *www.fao.org/publications/sofi/en/*.
2. The concept of triple crisis has been used in the past with different meanings; for instance, it has been referred to climate change, peak oil (the end of cheap energy), global resource depletion and extinction (see *www.ifg.org/events/triple_crisis_speakers.pdf*).
3. In fact, the global economy is passing through a period of dramatic change. The recent concern is with the financial crisis,

originating in the developed economies (first crisis). At the same time, climate change problem remains unsolved, with the growth in greenhouse gas emissions exceeding previous estimates (second crisis). Finally, malnutrition and hunger are on the rise, propelled by the present inflation in global food prices (third crisis). These three crises interact and could undermine the prosperity of present and future generations. Concerted actions are then needed in particular from international institutions.

4. See the debate about the opinion of the Nobel Prize winner, Paul Krugman, for instance at *http://blogs.telegraph.co.uk/finance/jeremywarner/100008311/will-someone-please-shut-krugman-up/*).
5. *www.un.org/esa/dsd/agenda21/res_agenda21_04.shtml.*
6. *www.unep.fr/scp/marrakech/about.htm.*
7. Sectors such as clean technologies, renewable energies, water services, green transportation, waste management, green buildings and sustainable agriculture and forests. More information is available at *www.unep.org/greeneconomy.*
8. A nice example to help in understanding this point is often quoted. When the Maasai people perform their culture for busloads of first-world visitors, they are not permitted to wear their digital watches, T-shirts, or football socks and all technological equipment is hidden from the tourists' view.

2

The meaning of tradition

Abstract: The aim of this chapter is twofold. First, it presents the meaning of the word tradition, starting from its Latin origin and continuing with the explanation of the role of tradition in sociology. Second, following the concept of tradition as a handing-down process, the chapter shows how the transmission of knowledge from one generation to another may assume a very important role for the recombinant process which generates every kind of novelty.

Key words: tradition, recombination, creation.

This chapter aims to explain what tradition is and why tradition may assume a strategic relevance. The roots of the word tradition may be found in the Latin language, since it derives from *traditionem* (nom. *traditio*), meaning 'delivery, surrender, handing down'. A practical explication of the Latin tradition as a delivering and handing-down process may be found in the Roman regulatory framework. Indeed, in Roman regulation, tradition referred to the formal delivery of goods. Specifically, two distinct types of tradition may be distinguished:

(i) 'consensual tradition', as the physical delivery of assets;

(ii) 'actual or symbolic tradition', as the delivery of a symbol that would give the opportunity to exercise the ownership on a property.

This first definition has the purpose of clarifying the basic idea underlying tradition, thus revealing that tradition is used to deal with a process in which something passes from one person to another. In other words, 'tradition' may be described as the handing down of statement, beliefs, legends, customs and information, from generation to generation, especially by word of mouth or by practice. Another definition of tradition refers to belief, principle or way of acting which people in a particular society or group have continued to follow for a long time. In this last definition, it is possible to perceive an idea of fixity, persistence and endurance of cultural traits of a specific society. On the contrary, the previous two descriptions of tradition manifest themselves as more dynamic processes. Despite the differences among the definitions of tradition, all have two key and common elements: culture and its mechanism of passing from one generation to the next. Thus there seems to emerge a tight link between culture and tradition: thanks to tradition, the collection of knowledge and practices can be transmitted from one generation to another. What is delivered or handed down from one individual to another is the set of intellectual knowledge acquired through study and experience. Likewise, a society or group can deliver or hand down to its members a system of knowledge, skills or beliefs, social institutions, political and economic and artistic activities and demonstrations of spiritual and religious life.

The most important religions have always adopted systems by which they transmit through history and generations the message of their founders. Examples are Jewish Mishnah, Islamic Hadith and Catholic Tradition. Mishnah is the compilation of oral tradition in order to ensure the preservation of Oral Law passed by God to Moses on Mount Sinai. Hadith refers to reports of statements or actions of

Muhammad, or of his tacit approval of something said or done in his presence. One aspect of this assumes a particular interest: the isnad. This is the chain of narrators, each mentioning the one from whom they heard the Hadith, which documents the route by which the report has been transmitted. Finally, the catechism of Catholic Church-defined tradition is the living transmission through which 'the Church, in its doctrine, life and worship, perpetuates and transmits to every generation all that she herself is, all that she believes'. This kind of transmission is based on bishops, who receive from their predecessors, ultimately the apostles who lived with Jesus, the full and living Gospel. Indeed, the Catholic Church specifies that Tradition, coming from apostles, hands on Jesus' teaching and example, and what the Holy Spirit had taught and continues to teach.

Now that tradition has been defined, it is interesting to consider how this, which is considered to be a means of transmission of knowledge and practice, affects the present and modernity. The main questions we want to address in this book are:

1. Do the knowledge and practices handed down by tradition remain the same or do changes occur?
2. Is the past a useful and fruitful source from which to innovate?

Starting from a sociological point of view, scholars from various disciplines observed that distinctive cultural traits endure over long periods of time and continue to shape a society's political and economic performance. Inglehart and Baker (2000) verified that different societies follow different trajectories even when they are subjected to the same forces of economic development, because of specific factors, such as cultural heritage, that shape how a particular society develops. Furthermore, Weber (1958) argued that traditional religious values have an enduring influence on the institutions of society. Gusfield (1967: 355) considered the relationship between the content of tradition and modernisation as 'interaction leading to a fusion and mutual penetration'. According to the author, modernism and tradition complement each other, rather than acting as opposite forces. Moreover, Shils (1971: 122) underlined the strong ties that the future presents with the past when he wrote that 'all novelty is a modification of what has existed previously'. He thought that within change there is persistence and vice versa, therefore without these two features, innovation fades and previous conditions would be restored. Thereby, what emerges from these sociological contributions is the persistence of strong cultural values, the essential role of past things in the creation of new, as well as the continuous process of modification and the mutual influence of old and new.

How is it possible? The phenomenon that best describes the permanence of the past in the present relies on the recombination of existing elements in order to create something of the new. This idea of recombination is stressed by the economic literature. In particular, Schumpeter (1934, 1939) affirmed that innovation combines old components in a new way, or that it consists in carrying out new combinations. Moreover, Nelson and Winter (1982) stated that novelty in art, science, or practical life consists to a substantial extent of a recombination of conceptual and physical materials that were previously in existence. A clear example of such a recombination process in art is provided by Picasso. Specifically, if we consider 'Les Damoiselles d'Avignon', it is the result of months of preparation and revision, combining elements as diverse as Iberian sculpture, African tribal masks and El Greco's painting to make this startling composition. Another interesting example of combination between very different elements that generates something new is represented by the 'Concerto for Group and Orchestra'. On 24th September 1969, the Royal Albert Hall in London was the stage of the first experiment mixing hard rock and orchestra. On this occasion, the rock band Deep Purple and the Royal Philharmonic Orchestra conducted by Sir Malcolm Arnold played an opera composed by Jon Lord, founder of the rock band. The result was a 'fusion' in which the instruments of orchestra, electric guitars and drums executed the same opera.

In the context of recombination, which usually occurs between components that are salient, proximal and available for the inventor/artist (Fleming, 2001), a knowledge store and a mechanism for the transmission of elements over time is very important. Along these lines, Eliot (1919: 1) thought of tradition as the 'the whole of the literature of Europe from Homer'. Hence, tradition becomes the enormous archive upon which to build the future. At this point, that thing that assumes a fundamental role is the artistic talent. This talent is the ability to recombine the elements of the archive so as to produce new relationships. For this reason, a single artist can create something only by what he/she knows and can create new combinations of past elements if he/she has a deep consciousness of the past. Sennett (2008) provided an interesting study of the relevance of tradition by analysing the dynamics of ancient workshops. In these places, knowledge is transmitted by means of assimilation through exercise, copying models and observing the movements and practices of masters. In the Middle Ages, guilds and corporations drew their economic power because they possessed a capital of knowledge transmitted from generation to generation, and because they were constituted by masters available to transmit this knowledge stock to other candidates. The source of economic power was the knowledge capital absorbed after years of work and delivered to other generations. The trials the apprentice had to overcome to become a master, make clear the process of the transmission and assimilation of knowledge. After seven years, the apprentice ran a *chef d'oevre*, a masterpiece, demonstrating the basic skills acquired. The implementation of the master was based on imitation, i.e. learned by copying. After a further period of four to ten years, he had to prove he deserved the title of master with a *chef d'oevre élevé*. At that time, the first test was intended to demonstrate the learning of processes and techniques, whereas the execution of a *chef d'oevre élevé* had the scope to demonstrate whether the craftsman was able to create something new and worthwhile.

In particular, the career of a goldsmith showed the typical elements of such a professional growth. The goldsmith apprentice was linked to his master's workshop for all the time necessary to learn to blend, purify and weigh the precious metals; all activities that required the practical and direct teaching of the master. Once he had presented, and had accepted, his masterpiece in the place of his apprenticeship, the young man was free to move from city to city to pursue his trade. In this way, he could watch new processes, new techniques, new tastes and could blend all these new elements with the knowledge already gained. At the end of this period, the itinerant jeweller presented his masterpiece to the college of masters. The itinerant jewellers had an added value, that consisted of all those experiences thanks to their movements and their exploration of the world from their first workshop, which they expressed in their handwork.

The experience of Matisse is very interesting for explaining the role of the workshop and originality. Matisse completed his artistic education when he was 27 years old after studying under the Symbolist painter Gustave Moreau, who sent his pupils to make copies of Old Masters paintings in the museum of the Louvre. This traditional approach encouraged students to focus more on methods of construction than on inventing new compositions. After this period, Matisse started to work independently, reusing and reworking his tools and techniques, and created his own style which featured forms with 'a truer, more essential character'. Similarly, the phenomenon of recombination may be found in technological innovation. In this regard, Arthur (2009) argued that novel technologies must somehow arise from fresh combinations of what already exists. This happens if there is a mechanism of 'heredity'. This means that novelty must descend in some way from the technologies that preceded them. A significant example is represented by the Video tape Recorder. This device combined previous systems and mechanisms, such as lateral scanning, with rotating heads and wideband video recording developed during the 1950s. This combination was due to the use of the knowledge of new mathematical analysis linked to the VTR FM system linked with the transmission of television images using an FM system, developed in the last years of the 1940s (IIT Research Institute, 1968).

Eliot's definition of tradition, together with workshops and the concept of heredity, helps us to understand how tradition affects the present. First, it clearly indicates the need to acquire existing knowledge that provides building blocks for creating something new. However, we strongly believe that what allows us to better understand why the past affects the present is the intuition that novelty is the result of a recombination of existing elements. In this way, the past continues in the present in the different ways in which an artist's or inventor's talent can elaborate. A further explanation of the recombinant view is provided by the metaphor of Bernard of Chartres (d. 1124). In particular, he used to say that we are like dwarfs who stand on the shoulders of giants, so that we can see farther than them not because of our stature or acuity of our sight, but because, standing on their shoulders, we are higher than them. Therefore, the giants represent the whole experience delivered by tradition, which allows novelty to be created. The meaning of this image may be clearer thanks to Ogburn (1922), who wrote about invention built cumulatively from earlier invention. Indeed, if new technologies are combinations of previous ones, then the stock of existing technologies must somehow provide the parts for combination. Primitive societies could not invent our modern technologies, since they did not possess the necessary ingredients and knowledge of how to work with them. The recombinant view of innovation also shapes a firm's behaviour. Nelson and Winter (1982) provided their interpretations about a firm's attitude in relation to knowledge recombination. At first, they proposed the routinisation of activity as the most important form of storage of the organisation's specific operational knowledge. The result is represented by the organisation's routines, considered as an abstract 'way of doing things'. So, routines may be seen as a means of embedding organisational memory and preserving the organisation's knowledge base. At the same time, routines become a useful building block for future new combinations. Therefore, firms draw from their own routines and they may be expected to behave in the future according to their past routines.

What has been said so far allows us to have a clearer idea of the role of the past, and then of tradition, in the innovation process. The current inspiration of novelty results from the combination of known elements into a new way. Tradition permits us to transfer over time the elements needed to innovate. This transmission may be operated by socialisation, teaching processes or personal efforts of learning. The main reason behind the increasing interest in innovation relies on the fact that it allows firms to be competitive and sustain such competitiveness. Nevertheless, among all of the multiple and possibly new combinations of elements, there are some of them that fit the environmental characteristics and thus succeed. This is the selection of what is better in a particular moment and in a particular environment. Hence, firms and inventors try to innovate in order to respond to market conditions and the best combination is chosen by the external environment. Following the market and environment, a firm's behaviour is influenced by the broader economic conditions surrounding them (Nelson, 1991).

When considering a specific technology, it is necessary to understand the existence of many variants developed by different inventors and diffused in different environments. Some technologies perform better than others, since they better address specific market needs and, consequently, they are selected for further use and development. The result is that firms act differently – and successful firms are able to operate following the differences and peculiarities of context in which they operate. Thus, firms are interested in knowing how to create something that fits the environmental needs. The same interest leads researchers to better understand the components involved in the innovation process. In this regard, some scholars have found that the old may sometimes play a key role in sustaining competition. Specifically, Fleming (2001) and Nerkar (2003) also demonstrated the need to experiment with old components in order to develop technological breakthroughs. This suggests that knowledge search and creation strategies have to be balanced between the past and the present, in order to effectively influence the future. However, according to Katila (2002), not all old knowledge positively affects the innovation process, since whether the age of knowledge hurts or promotes innovation depends on where firms search – close to or distant from their industrial sectors; within or outside their organisational boundaries; as well close to or far from their geographical areas. In addition, Katila and Ahuja (2002) argued that firms can create new and unique solutions not only by exploring new things, but also by mastering the old ones. Then, extant literature assumes the importance that old components may have for the innovation process. Once the important role that may be played by old components is recognised, the need to preserve and transfer such components to the future emerges. In this regard, tradition acts as a delivery process ensuring the transmission of old components upon which novelty is built.

In various markets many examples of products whose development has required the combination and exploitation of old elements may be found. For example, the TAG Heuer watch lets us remember the 1930s original models, but instead of a mechanical escapement there is now a solar powered microchip (Haynes, 1998). In addition, the 'new' Volkswagen Beetle is the result of the blending between the distinctive bubble shape of the old VW Bug and the latest automotive technology to produce a twenty-first-century car with twentieth-century styling (*The Economist*, 1998). Other examples in the automotive industry include the Chrysler PT Cruiser, a pastiche of the upright sedans of the 1940s; the BMW Z07, which evokes the two-seater Ferraris of the 1950s; and the 'new' Ford T-Bird, an affectionate nod to the style (if not the under-braked, overpowered performance) of the 1954 original (Hutton, 1999). Even more famous is the decision of Coca-Cola to bring back the moulded glass bottle that has become the symbol of this soft drink.

This decision leads us to discuss the case of Caffrey's Irish Ale. Launched in 1994, Caffrey's is one of the most remarkable marketing triumphs of recent years. A winning combination of the best features of lager (light, cool, refreshing) and more traditional Irish beers (mellow flavour, creamy head, slow to settle), the product integrates old and new in a strikingly original manner (Brown, 1999). But the significance of this case remains in the use of the past for promotion. The beer was launched on St. Patrick's Day, a national day in Ireland, and the television advert intimated the experience that you have during the drinking of this beer, aimed at reproducing the sensation of living in a forgotten Emerald Isle. In the food industry, there are many other examples of products presenting traditional features, as recognised by the Italian Ministry of Agriculture. Specifically, the ministry defines a traditional food product as an 'agrifood product whose methods of processing, storage and ripening are consolidated with time according to uniform and constant local use' (Ministero dell'Agricoltura, 1999: 1). Moreover, a recent study conducted by Guerrero (2009) tried to sum up a traditional food product from the consumers' perspective. The result was a product 'frequently consumed or associated with specific celebrations and/or seasons, normally transmitted from one generation to another, made accurately in a specific way according to the gastronomic heritage, with little or no processing/manipulation, distinguished and known because of its sensory properties and associated with a certain local area, region or country'. An interesting example is provided by the Pasta Benedetto Cavalieri, produced by old-world artisanal methods. Founded in 1918, the firm continues the family tradition of making pasta according to the time-honoured 'delicate method'. Hence, a traditional food product is essentially a handmade product typical of a culture and a territory that is transmitted from one generation to another preserving the same properties. These products have found a large number of consumers recognising the value embodied in such old elements. Thereby, the next question is why is there such a desire for the past?

This is the effect of the 'apprehension of the future' characteristic of societies in turmoil, those experiencing troubles, turbulence and transformation. A new survey from Roper Starch Worldwide shows that 55 per cent of Americans believe that the 'good old days' were better than today. This result is largely different to the attitude of a generation ago, when 54 per cent of Americans surveyed in 1974 told Roper they believed there was no time better than the present (Naughton and Vlasic, 1998). Examples of events that societies see as trauma without immediate solutions and shocks with uncertain outcomes are the dramatic collapse of Communism, the Gulf War, the 1992 European integration imbroglio and subsequent societal eruptions, the situation of late-colonial Hong Kong, the early years of the twenty-first century, the Islamic terrorist attacks, the emergence of ecological concerns, the emergence of new highly-competitive economic powers and the current financial crisis. All these events have created, and continue to create, favourable conditions for a developing retro mindset. This leads to an emphasis on conserving, restoring and protecting the past. The result is a broad request for cultural products that might construct a cultural identity based on a particular history before it (potentially) disappears, nostalgic films, the renovation of architectural landmarks, the popularity of old postcards, posters and publications, as well as the fashion for collecting antiques. These reconstructions offer the chance to step back in time to experience what life was like.

There is a desire to turn back the clock, fed by the sheer pace of contemporary technological change, and by the breathtaking onrush of the Information Age, with its high-speed modems, smartphones and pagers. While we benefitted from the wired 1990s, at the same time we are buffeted by the rapid pace of change. The futurist, Watts Wacker (Taylor, Wacker and Means, 1998), co-author of *The 500-Year Delta*, explains that 'we are creating a new culture and we don't know what's going to happen, so we need some warm fuzzies from our past'. Hence, the anxieties deriving from contemporary life create a form of escape and a desire for refuge in the experiences of the past. It may well have stimulated a yearning for simpler, less stressful times among those required to keep up, keep moving, keep ahead of the game. In terms of a social and philosophical point of view, justification used to explain the phenomena of present-day consumption, with its emphasis on nostalgic heritage products and experiences, may be found in the development of post-modern society. In particular, we are in a post-modernist era and post-modernism has key presuppositions that give explanations for retro orientation (Brown, 1995, 1997). The examples above confirm nostalgic inclinations, historical bricolage, a marked loss of faith in progressivist ideologies, and a preference toward the tried-and-tested over the new-and-improved. Signs of the past in a product may prove successful if they have an evocative power strong enough to give trust to customers, or if it has a nostalgic feature strong enough to be able to revive past experiences. Considering the above examples, the past may assume a strategic relevance when the old characteristics included in the product are highly distinctive of a firm, of a product, of a territory or of an historic period.

At the end of this chapter, which has ranged from etymology to religious culture, from artists' and artisans' workshops, to discography, from mechanisms of knowledge development to innovation generation, and from food products to cars, we are able to better understand the meaning of tradition. Specifically, tradition may be the process by which a firm selects distinctive components of its history, of a territory or of an age, and decides to transfer them over time in order to create perceived value for customers.

3

A tradition-based innovation strategy

Abstract: This chapter presents a firm innovation strategy based on the notion of tradition. Different types of tradition may be distinguished, such as that characterising a specific firm as well as a territory or an age. By strategically exploiting tradition, a firm may be able to innovate and differentiate itself from competitors, both by creating new value and appropriating it. In particular, creating value through tradition may be achieved by focusing on two distinct drivers – technology and design – applied at both product and process level. However, creating new value through tradition is not enough to assure firms of a sustainable competitive advantage. Thereby, to compete successfully through tradition, firms have not only to invest in the re-use and re-discovery of old knowledge and competencies, but also in the internalisation of all the intangible assets, such as values and culture, that contribute to the characterisation of the tradition. In other words, appropriating through tradition implies the creation of a tight relationship between products/processes and the identity of the firm, which has to reflect in its organisational routines the tradition it employs.

Key words: tradition, value creation, value appropriation.

In the current economic scenario, characterised by the recent financial crisis, the re-thinking of business models and the emerging of new customer needs, firms are re-discovering their roots, showing a renewed attention toward the notion of tradition. Examples of such behaviour may be found in different sectors, such as the automotive industry with the 'Mini' re-invented by BMW and the 'new' Beetle manufactured by Volkswagen, the pharmaceutical industry where an increasing number of companies are devoting their attention toward the re-use of natural remedies, and the food industry where the employment of traditional cooking ingredients and methods are more and more recognised and appreciated by customers. This growing interest in the past is also witnessed at the institutional level, as revealed by several initiatives aimed at sustaining and promoting the traditional values of geographical areas (see for instance the project funded by the Confindustria, the Italian Industrial Association, aimed at identifying and promoting the development of firms' historical archives and museums), as well as by the number of consulting firms that have seen a new business opportunity in the discovery of tradition. In this regard, it is important to mention The History Factory, a heritage management firm founded in Virginia in 1979 for helping to lead global corporations, organisations and institutions in discovering, preserving and leveraging their history to meet today's business challenges.

According to our own views, we define tradition as the whole set of competencies, knowledge, values and culture that characterises a specific firm, as well as a territory and an age. Thus, by combining hard and soft features, tradition allows firms to differ from competitors in terms of meaning and content. In fact, by exploiting tradition, firms may be able to innovate, thereby introducing new products and processes. The importance of traditional knowledge for innovation relies on the recombinant view (Gilfillan, 1935; Schumpeter, 1939; Nelson and Winter, 1982; Kogut and Zander, 1992; Rosenkopf and Almeida, 2003; Fleming and

Sorenson, 2004; Katila and Chen, 2008), which sees innovation as a re-combination of conceptual and physical materials that were previously in existence. For example, the computer workstation may be considered a novel combination of existing components including the CPU, motherboard, virtual storage and memory units, a display, a graphics processor, and systems and applications software (Fleming and Sorenson, 2004). Thereby, it emerges as a necessity for firms to search and explore over time, in order to find those old elements upon which to build novelty (Fleming, 2001; Katila, 2002; Nerkar, 2003).

Several factors can explain the relevance of the old. First, knowledge which has been prevalent in the market for a long time is generally considered more reliable (March, 1991), thus increasing the likelihood of creating successful innovations. In fact, older and more traditional knowledge can be subjected to more extensive testing over time and, consequently, firms may better realise the costs and benefits of using it, while decreasing the likelihood of errors and failures (Katila, 2002). Second, using older knowledge implies a decreased risk of retaliation from competitors relative to the use of more recent knowledge (Smith et al., 1991). The underlying assumption is that inventors usually initiate retaliation in response to the use of more recently created knowledge given that established knowledge tends to become available in the public domain. Hence, innovation that incorporates old knowledge faces fewer risks of misappropriation by competitors. Third, firms tend to build on recent knowledge which is more easily available. As new knowledge is created, it is typically promoted and disseminated in an attempt to reach prospective markets, thus simplifying the corresponding search process. In turn, older knowledge is less commonly sponsored and more difficult to find (Argote, 1999). In addition, recombining knowledge from broad time periods may be relevant thanks to the existence of complementary knowledge, institutions, or standards that significantly contribute to enhancing its value, thus allowing inventors who look back across time to find valuable solutions (Nerkar, 2003). Nevertheless, this doesn't necessarily mean that all old components are important, but that firms can differentiate themselves not only as to the extent to which they explore new things, but also as to the extent to which they search for and master the old ones (Katila and Ahuja, 2002). It is important to underline that knowledge creation strategies have to be balanced between the past and the present, combining and merging both new and old elements. For example, this is the case of mechatronics, which originated at the end of the 1970s with the fusion of older mechanical technologies and the newer electronics field (Freddi, 2009).

Following the main arguments of the resource-based view (Wernerfelt, 1984; Barney, 1991), it is possible to argue that tradition may constitute a distinctive resource for firms if they are able to translate it into organisational processes and routine. In fact, the sticky and embedded nature of tradition makes its imitation more difficult, thus contributing to its distinctiveness and rarity. Therefore, a strategic use of tradition may allow firms to create new value for customers, as well as to appropriate a great amount of it. Putting tradition at the centre of a business strategy implies the re-thinking of globalisation processes, shifting from the 'think global, act local' model (Kanter, 1995), which emphasises the importance of global best practices, to 'think local, act global', which aims at globally promoting local differences and peculiarities (Stone, 2001). The renewing of the past as a strategic approach is particularly relevant in the present post-recession scenario, where consumers are changing their needs and expectations, revealing an increasing nostalgia toward 'better days' and a demand for simplicity (Flatters and Willmott, 2009). Even prior to the recession, many consumers were starting to simplify. The US publisher Time Inc. recognised this trend early and capitalised on it by launching its highly successful back-to-basics magazine *Real Simple* in 2000. The recession has accelerated this maturing trend. A clear example is provided by the growing demand for trusted brands and traditional values, which certainly witness how consumers are shifting their interest toward products assuring a certain confidence about their quality.

Using tradition to create new products and differentiate existing ones may be achieved by focusing on two distinct drivers – technology and design. A tradition-based technology refers to the functionality given to products by old technological solutions and techniques that may be able to better address customer needs. In fact, even if new technologies present an improvement in current standards, they may leave unaddressed some elements of product value, such as performance, quality and usability – which may be the elements about which customers care significantly (Adner and Snow, 2010). Let us consider watches. Before 1969, all watches were powered by mechanical movement systems and a key performance indicator was accuracy. In 1969, quartz watches were introduced, offering an enormous increase in accuracy and a much lower cost. However, the rise of the quartz movement did not necessarily imply the shift toward the new technology, since watchmakers recognised that for the first time consumers could choose between mechanical and non-mechanical movements. Therefore, these watchmakers began to redesign their products beering in mind the segment of customers who valued the mechanical quality of a watch. Obviously, this segment was smaller than the broader watch market, but it is composed of consumers who were willing to pay a higher price. This is the strategy followed by companies such as Breguet, Rolex, Cartier and Vacheron Constantin, who focused on the traditional mechanical movement that made watches seem luxurious and distinguished objects.

The shift of old and traditional product technologies toward new markets or niches is also documented by programmable calculators and long-playing records. In particular, the former has been largely replaced by computers in business and science markets, but it now thrives in the education market, where the low price and portability make it perfect for teaching standard graphing concepts. The latter, despite the rise of new and better performing solutions such as CDs, continues to be sold to consumers given the better performance it may assure in terms of sound quality.

However, traditional product technologies may also compete with more recent ones. In the pharmaceutical industry, an increasing number of firms (GlaxoSmithKline, Pfizer, Merck & Co., AstraZeneca, Aventis, Bristol-Myers Squibb, Novartis, Pharmacia, Hoffman-La Roche and Johnson & Johnson) are developing medicinal products made from herbs and plants recognised as useful by popular tradition, as well as based on the recovery of application modes based on human history. An interesting example is represented by Aboca, an Italian firm founded in 1978 and located in the centre of Italy, in the area of Valtiberina Toscana, renowned since AD 1200 for the production of medicinal herbs. The company, operating in the agricultural, pharmaceutical and health fields, produces all of its products using natural herbs, which are directly cultivated, thus grounding its success on the symbiotic union between tradition and innovation.

The use of a tradition-based design at product level is related to the semantic dimension of products. In fact, product design is related to the meaning and values that it transfers to customers, for satisfying their emotional and socio-cultural needs (see Csikszentmihalyi and Rochberg-Halton, 1981; Margolin and Buchanan, 1995). In particular, it is associated with meaning and language, created by means of a combination of signs (i.e. indexes, symbols and icons) and forms (e.g. colour, material, surface, texture) that contribute to the symbolic value of products. Therefore, dealing with a traditional design means to create products that recall values and emotions belonging to the past. To better understand the relationship occurring between tradition and design consider some cases. The 'New Fiat 500' is an emblematic example of how Fiat Group decided to compete in the automotive market focusing not only on research and technological innovation, but also exploiting its own tradition, as well as that of the Italian socio-economic context of the 1960s. The car was designed respecting the concept of the original 'Fiat 500', launched in 1957, to embed into the new product all feelings and senses belonging to the first one, and that has contributed to make the Fiat 500 a masterpiece. This strategy has allowed the company to successfully distinguish its product from its competitors, giving it the capability to evoke in customers nostalgia for the 1960s and the economic boom they represent.

The use of a traditional design may be also found in other industries. The majority of the new coffeemakers introduced by Bialetti, including electrical ones, are designed following the original Moka created by Alfonso Bialetti in 1937, witnessing the important value recognised by customers for the traditional meaning of the product. After Coca-Cola recreated a plastic version of its famous contour bottle in 1994, sales grew by double digits in some markets, as revealed by Frank P. Bifulco, vice-president for marketing of Coca-Cola USA. Apreamare produces fishing boats according to the tradition of the 'gozzo sorrentino', as shown by its main characteristic, namely the double pointed hull, which is the same as the boat that was found a few years ago under the ashes of the ancient Roman town of Herculaneum.

Tradition-based technology and design can refer not only to final products, but also to production processes. In this case, the employment of tradition is closely related to the re-discovery of the artistic nature of production (Hall and Johnson, 2009). It generally involves the application of artisanal techniques and methods, as well as ways of organising production, that contribute to improving products and process quality, and to reducing their standardisation and augmenting the differences and the value that each customer may obtain. In fact, producing by using tradition may significantly increase output variety, thus promoting product distinctiveness and uniqueness. In a certain sense, firms that exploit tradition at production level can be compared to ancient workshops, in which craftsmen manufactured products that differed from one another and that were characterised by high quality, thanks to their skills and ability.

Artistic processes are often required where no consistent definition of quality exists. Therefore, customer value must be created by artists who devote considerable effort to understanding the individual customer preferences. The re-discovery of the artistic and artisanal nature of production coincides also with the resurrection of the 'homo faber' and the emergence of new business models founded on the pursuit of quality. As outlined by the sociologist Richard Sennet (2008), we are finding it necessary to re-evaluate the role of the artisan, as well as to devote new attention to cultural models that point to the central practical competencies and capabilities, thus recovering those values that have contributed to the development of the 'doing' culture, giving life to figures such as Cellini and Stradivari. This is the case of Steinway & Sons, producing pianos played by the majority of the world's concert pianists, who promoted the differences expressed by each piano as an indication of the richness of the materials and the craftsmanship that go into its products. In fact, Steinway & Sons knows that each of its concert grand pianos expresses a different personality and the company promotes that as a positive and distinctive characteristic. Likewise, master winemakers know that their job is to make the most of the distinctive qualities of each year's harvest. Still, Fabriano, a market leader in the sectors of office paper, drawing paper and security paper, continues with the manual production of paper following techniques from 700 years ago, thus emphasising the traditional skills and knowledge of workers and creating pieces and collections of high value.

Persol, a historical brand of the Luxottica Group, creates glasses by still employing manual techniques. Specifically, the glasses' flexible stems, called Meflecto, and the symbol of Persol, the Supreme Arrow, are manually created. As revealed by Fabio d'Angelantonio, the Chief Marketing Officer of Luxottica: 'Persol embodies the traditional values of the Italian hand-made art, thus becoming a world reference among luxury accessories.' Shifman, founded in Newark in 1893, produces mattresses and boxsprings using only artisanal and manual techniques, thus providing the company with a tradition of excellence in bedding that has endured for generations. Finally, the Real Paint & Varnish Company manufactures paints and colours following the tradition of craftsmen painters from before the Industrial Revolution, so that the repair and restoration of old buildings preserves the beautiful finishes of the past.

On the basis of the traditional drivers employed, i.e. technology and design, as well as of its application, i.e. product and process, it is possible to define a matrix

Figure 3.1 **Creating value through tradition**

	Product	Process
Design	Product-based design tradition	Process-based design tradition
Technology	Product-based technology tradition	Process-based technology tradition

explaining how a firm exploits tradition and at which level it works, respectively (see Figure 3.1).

However, creating new value through tradition is not enough to assure firms of a sustainable competitive advantage. In fact, as pointed out by several scholars (e.g. Teece, 1986; Winter, 2006), this strongly depends on appropriability, which refers to the degree to which a firm captures the value created when it introduces something new. Specifically, appropriability is related to the extent to which firms can limit other firms from imitation. Greater appropriability, modelled as lower spillovers from high-quality, low-cost firms to less efficient competitors, is associated with greater marketing power for the innovator, and it translates into larger cost differentials for the innovator, enhancing an innovator's ability to obtain higher price–cost margins and profits for the final product.

A lack of appropriability may explain why firms often fail to capture significant economic returns from the introduction of new products or processes. Many examples of this phenomenon may be provided (Teece, 1986; Pisano, 2006) and the EMI CAT scanner is a classic case. In the late-1960s,

Godfrey Houndsfield, an EMI senior research engineer, was engaged in pattern recognition research which resulted in his displaying a scan of a pig's brain. Subsequent clinical work established that computerised axial tomography (CAT) was viable for generating cross-sectional views of the human body – the greatest advance in radiology since the discovery of X-rays in 1895. However, while EMI was initially successful with its CAT scanner, within six years of its introduction into the US in 1973 the company had lost market leadership and by the eighth year had dropped out of the CAT scanner business. Other examples are represented by RC Cola, a small beverage company that was the first to introduce cola in a can and the first to introduce diet cola, but lost the market to competitors since both Coca Cola and Pepsi followed almost immediately and deprived RC of any significant advantage from its innovation.

Further, Bowmar, which introduced the pocket calculator, was not able to withstand competition from Texas Instruments, Hewlett Packard and others, and went out of business. More recently, Apple invented the graphical user interface, but Microsoft Windows dominates the PC market. Apple invented the PDA (the bricklike Newton) but Palm became the dominant player. Netscape invented the browsers, but Microsoft captured the market. In the medical field, Merck was a pioneer in cholesterol-lowering drug, Zocor, but Pfizer, a late entrant, grabbed the dominant market position with Lipitor. Excite and Lycos were the first real web search engines, but they lost out to Yahoo! and then to Google.

Therefore, both value creation and appropriation shape the competitive advantage, since value creation influences the potential magnitude of the advantage, whereas value appropriation influences the amount of the created value captured by the firm. In sum, value creation and appropriation both complement and serve as imperfect substitutes for each other (Mizik and Jacobson, 2003), allowing firms to achieve abnormal economic profits and competitive advantage.

Following the capabilities-based approach (Pisano and Teece, 1994; Teece et al., 1997), we would suggest that firms may also strategically use tradition to increase appropriability. In particular, we believe that to compete successfully through tradition, firms have not only to invest in the re-use and re-discovery of old knowledge and competencies, but also in the internalisation of all the intangible assets, such as values and culture, that contribute to characterise the tradition. In other words, appropriating through tradition implies the creation of a tight relationship between products (processes) and the identity of the firm, which has to reflect within its organisational routines the tradition it employs. By reinforcing its identity, a firm may be able to restrict imitations and to share with customers the traditional values and meanings that contribute to distinguishing it from its competitors. Several initiatives may be implemented to achieve this purpose. An important role is played by museums and historical archives, as revealed by the increasing number of firms that are paying more and more attention to rebuilding and communicating the fundamental steps of their history. Museums and archives allow firms to show the continuity between past and present, as well as the numerous changes and challenges they faced, to witness business strength and credibility. Interesting examples are represented by international brands such as Ferrari, Coca-Cola, Ducati and Guinness, which have created areas where customers may re-live the companies' history, understanding those peculiarities that contribute to make their products better than others, through the collection of characteristic objects, documents, images and movies.

As previously stated, firms may create customer value by employing not only their own tradition, but a tradition that is typical of a territory or, more generally, of a specific age.

Thereby, appropriability becomes more difficult since other companies may use the knowledge and competencies that shape tradition. Nevertheless, firms may strategically use tradition to capture the created value in this case too. Consider Barilla, the worldwide leader in the pasta business. The company creates value mainly by exploiting methods and techniques typical of the Italian food tradition. However, despite the fact that this know-how is not Barilla's property, the company has been able to transfer to consumers the uniqueness of its products. In this respect, a special mention has to be made of the Academia Barilla, the first international centre dedicated to the worldwide diffusion of Italian gastronomic culture, as well as to the prevention of imitations and falsifications. The Academia is located in the area of the first Barilla pasta factory and offers several services, all aimed at promoting and valorising the knowledge of Italian food quality. Therefore, significantly contributing to the protection of the Italian food culture, Barilla has been able to capture this tradition, appropriating the results deriving from its employment.

A similar strategic choice has been adopted by Perugina with its 'School of Chocolate', where people may learn the origins, history and features of chocolate, thus creating a tight connection between the firm and the Maya tradition of chocolate. Tong Ran Tang, a Chinese pharmaceutical group established in 1702, founded a hospital in Beijing to spread Traditional Chinese Medicine (TCM) culture, thus strengthening its tie with TCM as its main value-creation driver. Guinness, the world-famous Irish beer, continuously exploits the Irish tradition in order to make the product unique. The most fascinating initiative is represented by the Guinness Storehouse, where the company celebrates its tight link with Irish culture using different initiatives, such as the museum, the St. Patrick's Day Festival, the offering of traditional Irish food and drinks together with the Guinness beer, and the performance of old and new Irish music and artists.

Another interesting example of a firm exploiting an external tradition is represented by Brunello Cucinelli, an Italian company leader in the production of cashmere knitwear, which has developed its business following the traditions of a specific age. As revealed by its founder, Brunello Cucinelli, the firm may be defined as a humanistic company. In fact, the business has been developed following the tradition of humanism, which has significantly contributed to making this firm and its products unique. Such a humanistic approach is revealed by the great relevance given to employees and their necessities, as well as by the company's attention toward art and culture (about 25 per cent of profit is annually invested in cultural and artistic initiatives). This attention is physically set in the creation of the Forum of Arts Foundation, which is located in Solomeo and it is a system made up by various symbolically and physically correlated elements, representing a replica of the Roman interpretation of earlier Greek humanistic models.

Analysing the extent to which firms exploit tradition as a resource to create and appropriate value (see Figure 3.2), it is possible to identify:

- tradition-based firms, where tradition is used for both creating and capturing value;
- artisanal firms, where tradition is present only as a resource embedded into products and/or processes in terms of technology and/or design;
- old-fashioned firms, where tradition is used only as a marketing strategy, without being embedded in its products and/or processes.

Figure 3.2 **The tradition matrix**

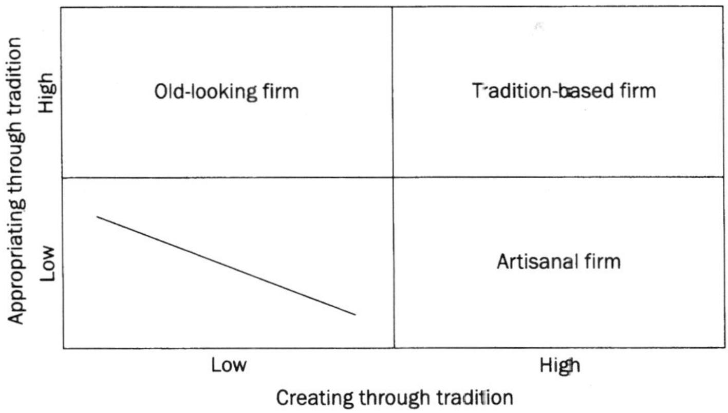

In this book, we are interested in analysing tradition-based firms. Therefore, in the following chapter we discuss a number of actual cases showing how companies may exploit tradition as a source of competitive advantage, blending tradition and innovation in both products and processes, and thus implementing a strategy that sees tradition as an effective opportunity for creating new value for customers and generating significant income from it.

4

Case studies

Abstract: This chapter describes and discusses the tradition-based strategies implemented by several companies. Cases have been selected following three main criteria: (i) a presence on the international market, (ii) the success of the firm in their specific industry, (iii) a high commitment toward craftsmanship, or toward the culture of a territory or an age. Specifically, the following companies are analysed: Aboca, Amarelli, Apreamare, Barilla Group, Breguet, Brunello Cucinelli Group, Caffè Vergnano, Cartier, Cartiere Milani Fabriano, Ditta Bortolo Nardini, Kikkoman, Lavazza, Lego Group, Lindt, Persol, Perugina, Pictet & Cie, Shifman, Steinway & Sons and Tong Ren Tang. For each company, a brief description is provided and then an analysis of the strategic use of tradition, in terms of both creation and appropriation, is presented.

Key words: international companies, tradition, strategy.

The study is based on a multiple-case design, which follows replication logic and allows a thorough understanding of the strategic exploitation of tradition. Seventeen firms are analysed as the result of a selective approach based on the following criteria:

- a presence on the international market;
- the success of the firm in their specific industry;

- a high commitment toward craftsmanship, or toward the culture and values of a territory or an age.

Cases have been selected from different industrial sectors and without geographical boundaries, thus providing grounded, accurate and general results. In addition, such a variation across firms allows the identification of both elements that arise in all or almost all firms and variable elements that explain differences among organisational processes (Sutton, 1984).

Aboca

Aboca operates in the natural health industry and is engaged in the development of herbal and cosmetic products sourced from medicinal plants. Specifically, it is simultaneously an agricultural, pharmaceutical and health company. Aboca adopts a process of vertical production, which allows direct control of every step of the production process: from planting to packaging of the finished product, all the way to its distribution to medical practitioners, pharmacies and herbalists. The company was founded in 1978 and is located in the centre of Italy, in the area of Valtiberina Toscana, known since AD 1200 for the production of medicinal herbs. According to its etymology, the term Aboca comes from the cleansing properties of the medicinal herb Camepizio, which was widely used in antiquity and which, in the Tuscan dialect, came to be called Abiga. The company's facilities include two centres of production with a total of 170,000 square feet, each equipped with the latest technologies; more than 1,700 acres of organic cultivation; a specialised chemistry laboratory and a staff of 300 employees. All of this along with the ownership of European process patents, makes

Aboca Italy's business leader in the natural health products sector, both in terms of know-how and sales. The success that the company has achieved is due to the value of its products, a symbol of the symbiotic union of tradition, research and innovation, conducted with dedication to a continuing ideal of promoting natural health care. In fact, given the recent rediscovery of the validity and usefulness of medicinal plants, Aboca invests considerably in the verification of the medicinal properties of plants as described by tradition, the search for new plants that are useful for human well-being and the improvements in the quality of those plants already being used.

The use of tradition in Aboca as a value-creation strategy is much evident in the whole production process, which is based on the adoption of medicinal herbs recognised as useful by popular tradition, as well as on the recovery of application modes grounded in human history. The company may well be considered as an organic farm, covering approximately 600 hectares, some of which are owned and some leased, throughout different areas of Tuscan and Umbrian Valtiberina. Approximately 60 species of medicinal herbs are cultivated on these lands, while another 20 species are wild-crafted by the employees. The agricultural phase begins with the selection of seeds and plants with a view to their propagation and multiplication, which, if necessary, can take place in specially-created and protected environments. A fundamental agronomic goal is to obtain highly therapeutic medicinal herbs through the selection of specific varieties with the greatest content of active ingredients. Following the selection of genetically-interesting varieties, the R&D department studies the chemical properties of the varieties in question and thoroughly researches their various possible medicinal applications. The entire agricultural sector follows the norms of GAP (good agricultural practice), which allows for a clearer designation of the raw material.

The guarantee of product quality is also provided by certificates of analysis with which all the specific data and properties of a predetermined species of medicinal herb are communicated. Throughout all the processing phases, Aboca aims to maintain the integrity and concentration of the active constituents present in the plant. Specifically, the company directly cultivates a large part of its own raw materials in order to guarantee an ample supply of high-quality raw materials, inspected and chosen on the basis of their biochemical features, and characterised by the absence of pesticide residues and other pollutants. Then, herbs are harvested employing both traditional and innovative techniques during the Saturation Period – that is the period during which the plant offers its best output in terms of the qualitative and quantitative worth of the phytocomplex. After the harvest, the fresh product is sent to a climate-controlled setting, where the quality control is carried out, in order to provide customers with products maintaining the integrity and concentration of the active constituents present in the plant.

Appropriating value through tradition is a well-consolidated strategy in Aboca, and it is shown in several initiatives aimed at linking the company with the tradition of natural herbs and its territory. Aboca always invests in the development and diffusion of a scientific culture of medicinal plants, financing scholarships in scientific research, organising courses and seminars for doctors, pharmacists and herbalists, and publishing books and articles, aimed at highlighting the tight blend between the tradition of herbs and scientific innovation. A tangible proof of such behaviour is also represented by the sponsoring of various university courses in herbalist techniques. Another important confirmation of this strategy may be found in Aboca's commitment to promote a law regulating the entire process from production to distribution for the protection of the consumer and the valorisation of traditional medicinal herbs. In addition, it is important to mention the Aboca Museum, which is the original and only Herb Museum that has rediscovered the history of the age-old relationship between man and herbs and is passing it down to future generations.

The Museum includes an educational section with the goal of spreading knowledge and understanding of its heritage in a multi-disciplinary manner. The Aboca Museum contains thousands of books preserved in a Bibliotheca Antiqua, which specialises in the historic theme of herbs and health over the centuries. Particular attention is given to the republishing of works written by illustrious figures of the Valtiberina area of Tuscany, thus reinforcing the tie between the company and its territory. In the Museum it is also possible to find art reproductions that evoke ancient emotions, such as medicinal ceramics, mortars and wooden boxes all made using traditional artisanal methods. Finally, the mission of developing new cosmetics and herbal products using traditional medicinal plants can also be seen in the company's name, derived from the medicinal herb known as 'Abiga' in Tuscan.

Amarelli

Amarelli is a historical family business established in 1731 in Rossano, in the region of Calabria in the south of Italy, for the cultivation, picking and production of liquorice. The Amarelli family arrived in Calabria from the north before the year 1000 and rapidly founded a vast domain, by uniting a large number of latifundia (spacious estates). Around 1500, the Amarelli, as well as all the large landowners, began the first harvest and sales of liquorice roots on their estates, in order to improve their land by letting it lie fallow before planting. In 1731, when Paolo Amarelli died at the age of 104, his son Fortunato and his grandson Paolo Fortunato created a 'concio', one of the first pre-industrial organisations, thus giving life to the Amarelli company. The enterprise continued to grow, passing from father to son, and in 1840 Domenico Amarelli opened a branch in Naples, the capital of Royalty and the second major European city after Paris. Then, in 1907, Nicola Amarelli innovated the production processes by installing two steam boilers, increasing the products' sales and appreciation. Currently, the company is managed by Pina Amarelli, who has significantly contributed to the international development of the company by adopting a communication policy mainly based on a dense network of relationships, both national and international, which combines the history and image of the company in order to raise awareness of the uniqueness of the product. In addition, to combat the competition which offers mass-produced and more affordable products, Pina Amarelli has focused the business on a niche market, turning to consumers wishing to capture the history, culture and craftsmanship that are typical of the Amarelli's production.

The distinctiveness of Amarelli relies on the craftsmanship of the production process, which has its roots in both the company and the territory's tradition. Since the foundation of the company, in 1731, the process is still the same. For example, a stone mould from the eighteenth century is still being used, the fuel which supplies the boiler is olive marc, and the final phase of the manufacturing, concentration, is still done under the supervision of the 'master liquorice maker'. The relevance of tradition in the Amarelli's business has been witnessed also by the Italian Chemical Company, which has assigned the company the Gold Medal, for having succeeded in uniting the excellence of the traditional craftsmanship of Amarelli products with the latest technology. Moreover, as argued by Pina Amarelli, the traditional values have always played a strategic role in the business development and success, since 'our ancestral know-how and the understanding of our business enables us to distinguish between the good ideas that are to be developed and those which have no chance of success even if, at the outset, they seem attractive'.

Another example of the strategic employment of tradition is evident at product-design level, i.e. in the famous little tins decorated with old pictures from the Amarelli's collection, where the liquorice is still presented. The introduction of these decorated tins goes back to 1931, when a child had an idea that he transformed into a nice sketch while he was tasting a little piece of liquorice. Thus, the child wrote a letter to the Amarelli baron, whose family had been producing liquorice for 200 years, suggesting he should create a decorated tin for the liquorice. The baron accepted the suggestion and so was born the pretty tin that today is still the symbol of the Amarelli liquorices. The tradition the firm exploits to create value, however, is not only internal. In fact, it is strictly related to the territory. According to the British Encyclopedia, Calabria grows the best liquorices in the world, especially on the Ionian Coast, where the microclimate enhances the content of glycyrrhizin, a sweetener which gives the distinctive flavour to the liquorice juice.

The valorisation of the Amarelli's tradition is guaranteed by several communication initiatives. The most important example is represented by the 'Giorgio Amarelli' museum, inaugurated in 2001. The museum was awarded the Guggenheim Enterprise and Culture Trophy and is located inside the historical residence of the Amarelli family which, since the late-fifteenth century, has always been in their possession. It presents old agricultural tools, a collection of clothing, reminders of the familial origin of the company, equipment and machines used in the making of the liquorice, historical documents, old images and commercial advertising. Furthermore, in honour of the Amarelli museum, the Italian Post Office has created, in the collection dedicated to the Italian Cultural and Artistic heritage, a stamp devoted to the 'Giorgio Amarelli' museum which had a print run of 3,500,000 stamps.

Apreamare

Apreamare is located in southern Italy and produces boats, yachts and fishing vessels. The company has a turnover of €19 million, 20 dealers throughout the world, and has 185 employees worldwide. The production plant in Torre Annunziata provides a boatyard measuring approximately 126,000 square metres, 30,000 of which are covered. There is an internal private dock which offers almost 100 moorings and a travel-lift able to be used on large-size boats.

Apreamare was founded in 1849 when Giovanni Aprea, shipwright, started to hand-build rowing and sailing boats used for fishing in Sorrento, basing his craft on the shape and form of the traditional 'gozzo sorrentino'. The entrepreneurial and manufacturing experience has been handed down from father to son. Indeed, the demonstration of this tradition is the role that Cataldo Aprea, one of the last descendants of Giovanni Aprea has in the firm. He is the president of this important business in southern Italy. It could not be otherwise, since he is a shipwright, a profession now uncommon, and because he is a supporter of the current corporate configuration. He began a new pattern with the introduction of a successful combination for the gozzo: motor propulsion in the 1930s. This is what has changed the history of Apreamare and the gozzo. From that time, Apreamare has concentrated on boats for pleasure-craft use, a field more prevalent than that of commercial fishing.

The Aprea family, leading this company, exploited the tradition of the region, thereby creating a boat that is a combination of the old 'gozzo sorrentino' and new solutions in terms of aesthetics and functionality. The key role played by tradition is also well authenticated by the boats' design. Its main characteristic, namely the double pointed hull with a very similar shape both in the bow and in the stern, is even the same as the boat that was found a few years ago under the ashes of the ancient Roman town of Ercolano. The pointed stern had a specific function: the fishing spot was reached by sailing and, once there, the boat was then pushed by stern-rowing, so that the stern became the workplace, as its particular structure guaranteed a better stability when the boat was still. The first design has remained largely unchanged since the years preceding the Second World War.

Apreamare has managed to 'navigate' in the wake of its centuries-old tradition and, at the same time, there have been constant design renewals and applications of all the latest technology to ensure performance and reliability for the most-demanding customers. The tradition also uses the capabilities of local artisans contributing to make its products' worldwide unique. The production process is still traditional, respecting the original characteristics of the gozzo, and it is mainly based on the skills of shipwrights, blacksmiths and carpenters. In particular, the manufacturing process follows an ancient method. To build a boat of seven metres, the artisans draw the first 'grace' from the originator, then seven ribs toward the stern and seven toward the bow.

In this way, the skeleton of the boat is formed. This phase is very important because the harmony and the hydrodynamics of the hull depend on the trace of the 'grace' and the shaping of the ribs. When the ribs are in place, the beams used to support the deck and the cockpit self-bailing are assembled. Then, the bulkheads are fitted out to divide the engine compartment, the forepeak and the cockpit. This step allows us to see the vessel in its full frame, ready to be coated with the plating. Beginning from the naked skeleton of the boat, the lining phase of the vessel can start. This is the moment at which the craftsmen start positioning planking for the deck (sunbathing), and the wood used is solid iroko. When the deck is completed, the ribs are lined by hot moulded plating. The boards are exposed to fire from inside and are constantly kept wet and under pressure. When the desired curvature is achieved, the boards are fixed, still hot, to the ribs with clamps. This ancient bending technique allows the boards to maintain their shape over time. Then, the long and meticulous task of moulding starts: the hull is cut long to eliminate any bumps or irregularities in the planting, which may otherwise negatively affect the hydrodynamics. Another important operation is the caulking (yield proof of the ship), placing the cotton cord between the intersections of the shell (seams). The caulking of the iroko deck is completed with a silicon sealant, a high-strength elastic material. After equipping the top edge with the gunwale, millings and stern bench, it is possible to start painting. The paint is a solid base with aluminum dissolved in oil and baked, which serves to nourish and lubricate the wood, and the hull is carefully filled and sanded before being painted again. After the dye base and filling, there is the final meticulous sanding. Finally, the decoration is very important for the craft and is carried out with great care, making a feature of the vessel.

Barilla Group

Barilla is today one of the top Italian food companies and has become popular worldwide as a symbol of Italian cooking, thanks to the attention to the quality of its products, as well as the significant investments in innovation and communication. By respecting its own traditional principles and values, which still feel current today, by considering employees as a fundamental asset and by developing leading-edge production systems, Barilla has become one of the world's most esteemed food companies – and one that is recognised worldwide as a symbol of Italian know-how. Barilla owns 50 production facilities (14 in Italy and 36 outside Italy) and exports to more than 150 countries. The company worldwide operates with different food brands, such as Barilla, Mulino Bianco, Voiello, Pavesi, Wasa, Harry's, Lieken Urkorn, Golden Toast, Kamps, Alixir, Misko, Filiz, Yemina and Vesta. Furthermore, the group includes the 'Number 1' brand, a group company engaged in logistics activities. Nowadays, the group employs more than 18,000 people and has net sales of more than €4.5 billion (2008).

The company has been managed for over 130 years by one family's entrepreneurial experience and is now run by the fourth generation siblings. Barilla was originally established in 1877 as a bread and pasta shop in Parma by its founder Pietro Barilla. In 1910, his sons Gualtiero and Riccardo took over the reins of the company, guiding the firm toward the industrial revolution and establishing the first pasta factory with 80 employees. During the 1970s, the company faced a financial crisis that forced the Barilla brothers to sell the majority share of their family business to the American multinational Grace. Nevertheless, in 1979 Pietro and Gianni Barilla, Riccardo's sons, turned an eight-year long dream into reality by buying back the family business, thus starting the re-launch of the company.

In a crowded field of over 2,000 Italian pasta manufacturers, Pietro and Gianni differentiated their company with a high-quality product supported by innovative marketing programmes. Barilla revolutionised the Italian pasta industry's marketing practices by creating a strong brand name and image for its pasta, selling pasta in a sealed cardboard box (1955) with a recognisable colour pattern, rather than in bulk, and investing in large-scale advertising programmes. After Pietro's death in 1993, leadership passed to 37-year-old Guido, the oldest of the three sons, with Luca and Paolo also taking on increased operating responsibility. Realising that growth would bring significant challenges, the family hired Edwin Artzt, former chairman and chief executive of the Procter & Gamble Company (P&G), to spearhead its drive overseas. Artzt joined in August 1995 as part-time executive director and quickly moved to upgrade the management, replacing most of the top team with individuals experienced in fast-moving consumer goods. He began an aggressive expansion outside of Italy, pushing distribution and national advertising programmes in order to rapidly increase market share. The Barilla brothers were involved in all major decisions and were members of three key committees: the executive committee, the brand equity committee and the product development committee. The family has always played the role of innovation as a driving force and, at the same time, guarantor of traditional values, thus assuring the safety of their products. The importance for the company to be pushed and driven by new products is also witnessed by the great effort Barilla spent in product research and consumer development. Specifically, research was conducted in-house in Parma, with over 2,000 people passing through the firm's 'Insight Barilla' facility every year. The centre included six food-tasting stations and two ethnographic kitchens, where consumers were observed and filmed as they cooked.

The central role played by tradition in Barilla's production processes may be noticed in several aspects and is evident in the mission statement of its founder Pietro Barilla: 'give to customers what you will give to your sons'. During its history, Barilla Group has shifted from an artisanal production to an industrial one, while still preserving the peculiarities of handcrafted products. A clear example of this is provided in the way in which they produce pasta, which follows a bronze-die shaping process and is strictly based on methods and techniques typical of Italian tradition. Moreover, the raw materials employed in the pasta production follow a tradition-based approach. For example, Barilla uses more than 500 million eggs, which are purchased only from producers able to internally manage the whole production cycle. The re-discovery of tradition to create value is also revealed by similar company products: recently, it has launched a new product line named 'Le Regionali', where Barilla re-introduces the different types of 'old pasta', characteristic of each Italian region.

As explained in the previous section, creating value through tradition is only the first step of a tradition-based strategy, since firms need to capture this value. In this regard, Barilla bases its appropriation strategy in creating a tight link between its products and Italian food tradition. Such a strategy is pursued following both communication initiatives and the valorisation and diffusion of the firm's history and Italian gastronomic heritage. Since 1910, Barilla has adopted a communication approach in relation to the artisanal tradition of pasta, focusing on the values of family and home-made cooking. According to company managers, Barilla was the strongest brand name in Italy: one-third of

Italians would name Barilla first when asked to name any brand that came to mind. Household penetration was over 80 per cent. The company spent approximately €100 million a year on advertising in Italy, primarily for the higher-margin bakery line. Historically, the company had used an emotional 'Barilla means family' message.

In 1987, Barilla created its historical archive to collect and preserve all the documents witnessing more than a century of the firm's life, thus contributing to strengthening the traditional content of its corporate identity. The most important investment made by Barilla in terms of appropriation strategy is represented by the foundation of the Academia Barilla. The genesis of the Academia began in 1998 with Guido Barilla's idea of constructing a training centre as part of an extensive development planned for the old Barilla pasta factory site in Parma. Formally stated, the Academia's three-part mission was to:

- preserve authentic Italian food products and protect them from imitation;
- develop and support the Italian food arts through culinary training that fostered innovation and creativity among chefs in Italy and abroad;
- promote and raise awareness of Italian gastronomic culture around the world through events, initiatives and publishing.

Moreover, the Academia commercialised an own-product line, initially focused on the handful of categories essential for Italian cuisine, such as regional olive oils, balsamic vinegars, cured meats and cheese (Parmigiano). Pasta was excluded to prevent competition with the core business of the Barilla Group. High-quality packaging was designed, featuring a map of Italy that showed the production region for each product. Differing from its parent, Academia did not produce anything itself. It added value by buying the finest products from the local producers and then doing the sales and marketing. Products were selected by local chefs and experts in regional gastronomy whose job was to taste and identify the best. The superior characteristics of these products depend on the traditional production techniques, which, handed down from generation to generation, were usually labour- and cost-intensive and, for products such as cured meats, cheese and balsamic vinegars, often involved a lengthy aging process. For example, the cows whose milk was used to produce Parmigiano cheese could only be fed with a certain type of hay, and that hay had to be stored naturally. For each wheel of Parmigiano, full-fat milk collected in the morning was blended with skimmed milk collected the evening before in a tub owned by the farmer. The master cheese-maker added only curd and salt. After hand-moulding, the wheels were labelled to show ownership, then dated and stored to age. They had to be turned often during storage to prevent spoiling, although some spoiled anyway. After 12 months, a representative from the local Consortium tested each wheel for quality by tapping it with a hammer. Solid-sounding wheels received the Consortium stamp and could be sold as Parmigiano. However, many wheels continued to age since the best cheese for grating was even older, between 18 to 24 months.

The Academia Brand was first introduced in the United States for three main reasons. First, Italian food was the second most popular choice after local cuisine. Second, there were as many Italian restaurants in the United States as there were in Italy. Third, the number of 'nonauthentic' Italian products sold confirmed the high potential of the market. In May 2004, the opening of the Academia Barilla Culinary Centre in Parma signalled the formal launch of the Academia and generated a wave of positive publicity in the United

States. The Centre contained 16 state-of-the-art kitchens, specialised laboratories for pastries and pizza, laboratories for tasting and sensory development, and a 90-seat auditorium for demonstrations by guest chefs. Fee-based classes for professional chefs and food lovers were conducted in both Italian and English. The Centre also housed the Academia's 7,500-volume gastronomic library, which included more than 3 million recipes going as far back as the sixteenth century. By mid-2006, the Academia products were served in 1,500 top US restaurants and were available at 500 gourmet food stores. US revenues for the year were projected to be $10 million and were split about evenly between food-service and retail. The sales breakdown by product was approximately 35 per cent olive oils, 29 per cent cured meats, 30 per cent cheese, 4 per cent balsamic vinegars and 2 per cent other products such as condiments. All the different activities undertaken by the Academia are aimed at promoting and valorising the knowledge of the quality of Italian food. By significantly contributing to the protection of the Italian food culture, Barilla has been able to capture this tradition, and to enjoy the advantages that it offers. In other words, thanks to the Academia's foundation, Barilla is able to market around the world the strong values of the Italian way of eating, both in taste and nutrition, thereby bringing the best of Italian culture and food to the world and becoming a sizeable and profitable business.

Breguet

La Montres Breguet is a limited company that is part of the Swatch Group, located in L'Abbaye, in the Vallée de Joux, in the West of Switzerland. The company employs between 50 and 100 people. Among its clients, Breguet includes several illustrious individuals, such as Queen Marie-Antoniette, Napoleon Bonaparte, Talleyrand, the Sultan of the Ottoman Empire, Caroline Murat, Tsar Alexander I of Russia, Queen Victoria, Sir Winston Churchill and Arthur Rubinstein, who put their confidence in the taste and artistry of the company. The list of estimable customers continues today, as revealed by a special list drawn up by the 2007 Business Montres & Joaillerie, which classifies Breguet at twelfth place among the 50 best watchmakers in the world, sorted by the value of their brand.

The history of the company began when its founder, Abraham-Louis Breguet, born in Neuchâtel in 1747, began a high quality watchmaking business, setting up on Quai de l'Horloge, Ile de la Cité in Paris. In a short time, Breguet showed his great innovative ability and kick-started his career with a series of breakthroughs. Specifically, in 1780 there was the launch of the first self-winding watch known as 'perpétuelles', with an oscillating weight and two going-barrels. This represented the beginning of a very rich and fruitful innovative period, characterised by a number of new technical solutions, such as the invention of the 'pare-chute' anti-shock device, the development of a perpetual date calendar and the Breguet balance-spring and ruby cylinder. Moreover, the genius of Breguet developed the 'tourbillion', created in 1795 but patented in 1801, which has become the most distinguishing feature of the mechanism by which Breguet watches mark the time. The tourbillion counters the effects of gravity by mounting the escapement and balance wheel in a rotating cage, ostensibly in order to negate the effect of gravity when the timepiece (and thus the escapement) is rotated. Originally an attempt to improve accuracy, tourbillions are still included in some expensive modern watches as a novelty and a demonstration of watchmaking virtuosity. At the end of the eighteenth century, the first single-hand watch, known as the subscription watch, was developed and produced by Breguet. The traditional values and competences characterising the company, as well as its success, has been preserved over the years, despite the change of ownership from the Breguet family to the Brown family and the acquisition of the company by the Swatch Group in 1999.

Integrating tradition and innovation is a specific feature of Breguet. The history of the firm clearly shows that the technological base on which Breguet continues its innovative effort is the result of patents filed by the founder. Specifically, the tourbillion is the main element of Breguet tradition, which is recombined, reused and improved many times over. Now, this mechanism has been made visible, in order to further characterise the watches and to increase their value. Despite the variations between watches, Breguet tends to implement processes using modern sophisticated equipment that complements the traditional process. The watchmakers employed by Breguet are among the most famous and important, capable of building, shaping and finishing each piece by hand. In fact, the common denominator of all Breguet's production process is the handwork. For example, evidence of the handmade production are the Arabic numerals of the Cuvette model, entirely made by hand. Moreover, the Tradition collection is composed of pieces created using the exact rules established two centuries ago, and the Classique line is made exactly by following the constructive traditions of Breguet. All these handmade watches respect as closely as possible the original traditions of the company, with manual or automatic loading complications.

The above traditional elements have given success and fame to Breguet watches but, at the same time, they have highlighted the problem of counterfeiting, which had already been of concern to Abraham-Louis Breguet. In particular, to protect its customers and combat imitations, Breguet has started to instigate unique characteristics – which have contributed to shaping the traditional value of its watches – such as engine-turned dials, Breguet numerals, lugs and caseband fluting. Since Breguet's early days, the manufacturing numbers of its watches which identify them for posterity, have enabled collectors to confirm their origin and provenance. In keeping with tradition, the unique production number assigned to each Breguet watch is used to testify to the talent and the care taken in its manufacture for generations to come. In 1795, Breguet added a further countermeasure to prevent imitation, namely the secret signature. Etched into the dial, the signature is all but invisible unless the dial is examined in oblique light.

Around 1786, Breguet began fitting his watches with engine-turned silver or gold dials of his own design. The craft of carving recurring patterns in metal is rare today, but it remains one of the details that identify Breguet watches. Manually engine-turned, their dials are celebrated for the fineness of their patterns, reflecting the regularity of the movements within. Breguet watches have featured the founder's celebrated hollow, eccentric 'moon' tip watch hands for over two centuries. Designed around 1783, his slim, sleek hands proved an instant success and the expression 'Breguet hands' soon became a common watchmaker's term. Simple and easy-to-read, they are found on most Breguet timepieces and have been widely imitated by others. Some Breguet watches display the distinctive numerals that Abraham-Louis Breguet designed. Although he himself was no calligraphist, Breguet's Arabic numerals show his flair for combining function with elegance. Still used today, particularly on watches with enamel dials, Breguet numerals first appeared before the French Revolution when they shared the dial with tiny stars to mark the minutes and stylised fleur-de-lys at five-minute intervals. By 1790, they had assumed their definitive form. Although essential only to wristwatches, the lugs that link the strap to the case bear all the hallmarks of authentic Breguet styling. Screw-pins, rather than the more usual sprung bars, hold the strap between the horns, a solution that is not only better looking but also more secure. The lugs have to be welded onto the caseband as much for technical reasons, such as rigidity and strength, as for aesthetic consistency.

All the above arguments help to explain how the innovations created by the firm are founded in its technological and cultural heritage, and continue to be a starting point and characteristic feature of each Breguet model. The strategic relevance played by such heritage highlights the necessity to protect it, as well as to communicate and share it with customers. In this regard, in 2000, the Breguet Museum was inaugurated by Mr. Nicolas G. Hayek. It presents invaluable documents and items relating to the history of the House of Breguet, founded in 1775. Substantially enlarged and with its collections augmented, the Museum is now located on the first floor of the Breguet Boutique at 6 Place Vendôme, in Paris. The Breguet archives cover over two centuries and include production registers, sales ledgers, repair books, certificates of authenticity, letters from clients and technical annotations written by Abraham-Louis Breguet and his son. To build up its current collection, the company is making new acquisitions every year. Today over 100 timepieces testify to the endlessly inventive mind of Abraham-Louis Breguet – as well as to the variety of his contributions to fine horology, all combining technical prowess with elegance of form. One can admire perpétuelles (automatic), souscription, tact, simple or repeating watches, as well as some of the oldest keyless watches, several travel clocks, marine chronometers and military watches. Another analogous initiative was undertaken by the company together with the State Hermitage Museum of St. Petersburg, to exhibit several Breguet antique watches. This collaboration originates from the long and profitable relationship between Abraham-Louis Breguet and the Russian ruling family.

Brunello Cucinelli Group

The company was founded in 1978 by the entrepreneur Brunello Cucinelli and is located in the medieval village of Solomeo, in the centre of Italy. The Group is an international leader in the production of cashmere knitwear and operates under the brands Brunello Cucinelli, Gunex and RivaMonti. Nowadays, the Group employs about 500 people, exports more than 60 per cent of its production, and had net sales of €144 million (2008). Recently, Cucinelli has won two prestigious awards – the 'Imprenditore Olivettiano' in Italy and the 'Best of the Best' in USA, for his ability to blend elegance and casual style in a single product.

The relationship between the Group and its use of tradition is clearly revealed by the artistic and artisanal nature of the production process, which is grounded on the competences and workmanship found in the textile industry of the Umbria region. Rather than being viewed as an industrial firm, Brunello Cucinelli may be considered as a haute couture, where a quality-control process is executed on each product and there is not an automatic system used either in the design or in the production phases. The Group internally manufactures only the base model for every product line; the other models are manufactured by external producers, which are constituted by artisanal tailor's shops. Cucinelli supplies these shops with the original model, together with designs and materials.

The Group captures the value created by exploiting two different types of tradition:

- the tradition of the area in which it is located;
- the tradition of a specific antique age, i.e. humanism.

Specifically, Brunello Cucinelli has created a perfect symbiosis between his firm and the region, as witnessed by the fact that the brand of the Group is represented by the ancient escutcheon of Solomeo village. Offices and production sites are located in old medieval buildings, completely restructured, while still maintaining and preserving their traditional characteristics. All the Cucinelli shops located around the world typically reflect the same style of Solomeo village.

However, the use of tradition is not to be considered only as a sterile recovery of old places. In fact, these places contribute by shaping the routines and practices by which the Group operates, thus embedding their tradition into the firm's DNA. In addition, the founder developed his business following humanistic values and tradition, thus giving life to what he names the 'Humanistic Company', which is defined as a company that responds in the noblest sense to all the ethical codes that humans have defined over the centuries. Such a humanistic approach is revealed by two main aspects. First, people and their needs are placed at the centre of any production process, as demonstrated by the lack of any formal controls, such as signs of authority. The great relevance and trust given to employees is fundamental to obtain the outstanding quality and beauty of an artisan product. Second, the humanistic nature of the Group is also much in evidence in Brunello Cucinelli's approach to patronage, as shown by his steady attention to art and culture (about 25 per cent of profit is annually invested in cultural and artistic initiatives), formalised in the creation of the Forum of Arts Foundation. The Forum of Arts is located in Solomeo and is a system made up of various symbolically and physically correlated elements, such as the amphitheatre, the academy, the library, the Renaissance-style theatre and the philosophy garden. As a whole, the complex was built to represent a centre of friendship and culture, a replica of the Roman interpretation of earlier Greek models.

Caffè Vergnano

The company, the oldest domestic coffee-manufacturing firm, was founded in 1882 in Chieri, a small town at the foot of the Turin hills, where it began as a small grocery shop, set up by the grandfather of the current owner of the company, Domenico Vergano. The 1930s signalled an important stage in the growth of the company, with the acquisition of a coffee-producing farm in Kenya and the opening of three warehouses in Italy located in Turin, Alba and Chieri. While the company continued to grow, it encountered further success in 1986 thanks to the merger with the historic Casa del Caffè in Turin, a high quality market leader in the bar/coffee shop sector of Piedmont, which now serves more than 2,000 clients throughout the region. Today, Caffè Vergnano exports its coffee to over 60 countries and employs a staff of 90. With an 8,000 square metre plant, located in Santena (Turin), 11 automated production lines, over 80 established products and double-digit growth since 2001, Caffè Vergnano finished 2008 with €50 million of preliminary sales in 2008 (the fiscal year closes on 30th June), broken up into 57 per cent within mass market channels, 26 per cent in the Hotel-Restaurant-Café sector, 15 per cent in exports and 2 per cent in franchising – thus confirming the brand to be among the principle players in the global coffee industry.

The success of Caffè Vergnano is based on the blend of two fundamentally distinctive traits – tradition and innovation. The coffee bean selection process still remains a family affair, as demonstrated by the fact that the president of the company himself, Carlo Vergnano, together with his son Pietro, visits the plantations and the suppliers in their respective countries of origin. During the roasting phase, which is the most important step in the coffee production process, the company still uses traditional methods and techniques. Specifically, the different coffee varieties are roasted separately, type by type, so that just the right level of roasting can be achieved. In contrast to the majority of modern, large roasting factories, which function according to specific 'turbo' industrial methods in 5–6 minute cycles at a high temperature, Caffè Vergnano uses a 'slow' roasting system, with cycles which last between 18 and 22 minutes. For each new cycle, the roasting of each batch is individually controlled, to guarantee a product with perfectly uniform quality, roasted for the correct amount of time.

Handcrafted production processes are supported by continuous innovations, especially in packaging. An important example, in this sense, is the PET plastic container for Miscela 1882 Espresso. The innovative design and the transparency of the container gives a visual way to recognise the product. In this way, the guaranteed quality beans selected for their colour and shape are left to be admired instead of being hidden from the eye.

In order to ensure the valorisation of its tradition, as well as the appropriation of the value that Caffè Vergnano creates by exploiting its well-established experience in coffee manufacture, the company is committed to two projects: 'Caffè Vergnano 1882' and 'Accademia Vergnano'. Caffè Vergnano 1882 is a project which oversees the development of coffee-bar networks in Italy and around the world, from

London to Hong Kong, from Milan to Paris, and even places in Saudi Arabia which are dedicated to the cult of coffee. The project deals with refined and elegant locations where it is possible to experience the traditional and unmistakable aromas of a perfect espresso. In the more than 40 locations that already exist, it is possible to enjoy the flavour and atmosphere of Italian coffee all the way through. Caffè Vergnano 1882 has been developed to allow customers to live a true, multi-sensory experience of coffee and of 'Made in Italy' in environments which present an atmosphere of elegance and refinement in design and aromas. From the flagship store to the corner stores within shopping centres, each location suggests its own concept, distinctively personalised and adapted to its surroundings. Just like the historic first Vergnano 1882 coffee shop in Chieri, all the others exhibit a squared, wooden storefront and are characterised by their refined interior essentials, such as the use of lights and the choice of natural materials like oak or slate. The peculiarity of the Caffè Vergnano 1882 project requires a particular selection of locations. To do this, the company is oriented toward locations which correspond to important parameters in regard to the capitals, or strategically important city centres, in the heart of the city and in the commercial areas, with the possibility to develop outdoor areas. The sales area should be a minimum of 50 square metres, like a traditional bar, with a glass surface of at least 8 square metres. These positions within an airport or a large commercial centre are very important. All of these characteristics allow for the implementation of easily recognisable, and therefore easily reachable, sites. Another element which the company takes into account is the integration of the site with the local architecture, to obtain a classy, yet personalised location within its surroundings.

Founded in 2005 as a school dedicated to the training of bartenders for the preparation of the espresso, the Accademia

Vergnano is becoming a point of reference in Italy and throughout the world with the objective of promoting the true culture of Italian coffee bars. This project assumes a strategic role for the company, since it sets the Accademia as the reference and meeting point between the long-lasting experience of the company regarding coffee and those who 'stand behind the bar', providing the consumer with perfect quality and a full service. The Accademia Vergnano provides for an articulated programme of different training activities, such as:

- *Coffee discovery*, which provides a brief introduction to the world of coffee from the plant to the cup, covering the Caffè Vergnano roasting process to giving an in-depth understanding of how to make the espresso.
- *Espresso for professionals*, a complete course which seeks to deeply understand the world of coffee, the diverse areas they come from, the cultivation and processing methods.
- *Latte art*, a course to develop modern manual coffee-making techniques.
- *Creative coffees*, which provides an advanced knowledge for the creation of espresso-based beverages combined with different products to formulate new drinks.

Cartier

The company was founded in 1847 in Paris, when Louis-Francois Cartier took over the jewellery workshop of his master and, within a short time, he decided to support the workshop where he created his own jewellery with a shop in which he could meet customers and sell his products. In 1874, his son, Alfred Cartier, took over the company, and his sons (Louis, Pierre and Jacques) in turn became responsible for the firm after him. Louis took charge of the Paris office located in the famous Rue de la Paix, in 1899, whereas Jacques took care of operations in London and created a subsidiary in New Bond Street. Finally, Pierre founded the most famous subsidiary in New York in 1909, and in 1917 moved into the well-known building on 5th Street. After the death of Jacques and Louis, which occurred almost simultaneously, Pierre Cartier headed the company until his death in 1965. Then, his daughters and son Claude dismantled the company. In 1972, the company was consolidated again and bought by a group of investors led by Joseph Kanou, and Robert Hocq, the creator of Les Must de Cartier jewellery, became the company president.

The creation of Cartier jewellery originates from the coming together of the jeweller and the stone, the craftsman and the precious metals. From the initial design to the final piece, all the phases involve the typical features of traditional jewellery, thus combining individual competencies and talents to create jewellery. Each piece of Cartier jewellery may be considered as a work of art, created by an impressive synergy of artistic skills and knowledge, going through all of the company's departments. Among the different types of Cartier products, particular attention has to be paid to watch production. Located in La Chaux-de-Fonds, at the heart of an area steeped in watchmaking tradition, the Cartier Manufacture is one of the largest fully-integrated production facilities in Switzerland and extends over an area of 30,000 square metres. All areas of watchmaking expertise are represented here, with 175 specialties divided into three major groups: development, production and customer services. Striking a subtle balance between cutting-edge technology and artisanal craftsmanship, the Cartier Manufacture above all depends on the outstanding expertise of over 1,000 employees from 37 countries. It features an incomparably rich blend of human resources and is the trustee of ancient know-how that Cartier is determined to preserve, both by internal training and the integration of artistic professions. It is commonly recognised that watch-making is a delicate art, since creating highly sensitive mechanisms calls for unique expertise combined with specialist manual skills. The Cartier Manufacture is thus strongly committed to preserving the essence of this ancient *savoir faire* from day to day, through activities that promote the transmission and teaching of these arts. For this reason, Cartier has set up a watchmaking school, l'Institut Horlogerie Cartier, to provide training in new skills that will ensure the future of Swiss watchmaking expertise. Similarly, the integration of artistic crafts within the Manufacture itself, such as enamelling and gem-setting, also demonstrates Cartier's determination to preserve these rare and thus vulnerable skills.

The value of tradition plays a key role in the successful strategy of Cartier, becoming the common denominator of a series of initiatives to support and reinforce the myth and the uniqueness of its products. These actions aim at creating a well-established and consolidated identity, which is then reflected into all of the company's activities. The Fondation Cartier pour l'Art Contemporain was founded in 1984, giving a redefinition of the concept of corporate patronage in France. Helping a wide variety of artists to invent, create and display their creations, the Fondation promotes contemporary art for its ability to give voice to the concerns of the modern world, thus making it more easily accessible to a wider audience. Sited since 1989 at the centre of Cartier's heritage policies, the Cartier Collection has always enjoyed a level of visibility that reflects its historical and stylistic importance. The collection has been exhibited in internationally renowned museums such as the Petit Palais (Paris), the Metropolitan

Teien Art Museum (Tokyo), the Metropolitan Museum of Art (New York), the Palacio de Bellas Artes (Mexico City), the National Museum (Shanghai) and, more recently in 2007, the Gulbenkian Foundation (Lisbon) and the Kremlin Museum (Moscow). The Cartier collection includes 1,300 pieces that illustrate the evolution of style and technique of Cartier creations. Each of these items, purchased by Cartier from private collections or at international auctions, has its own history and brings to light bygone eras, as well as commemorating the lives of famous people who owned the jewellery or commissioned their creation.

The search for and trade in such vintage creations are managed by the Cartier Tradition Department. The Cartier Archives, located in Paris, London and New York, classify all items produced by Cartier, thanks to a series of very accurate catalogues that track the daily lives of the company. The Paris collection includes about 40,000 negatives, 30,000 of them on glass plates with gelatin-bromide. The collection of plaster casts (also in Paris, including casts of the period between 1905 and 1915) offers a three-dimensional view of the workshops. The department dedicated to the archives contains an important bank of images, preparatory drawings, sketches and creations that exhaustively illustrate the process of jewellery creation. Another important initiative undertaken by Cartier is represented by the Fondation de la Haute Horlogerie, which is devoted to promoting and to developing the culture of technical and precious Haute Horlogerie in the world, linking art and technology.

Cartiere Miliani Fabriano

Since the thirteenth century, Cartiere Miliani Fabriano has been a historic, yet modern business with characteristics that are unique in the paper mill world. It now has a turnover of approximately €247 million (2009), has sold around 193,000 tons of paper and has four production plants and 626 employees. Despite operating in the exclusive production of wood-free uncoated paper, the company is present in the market for office paper, security paper and banknotes, drawing and fine arts paper, paperboard for printing and packaging, while also retaining a department where paper is still produced by hand. Production of the various different types of Fabriano paper takes place at four different plants: Fabriano, which employs a workforce of approximately 270 people and occupies a covered surface area of 36,000 square metres; Rocchetta, which is the newest production plant, which employs 112 people and occupies a total area of 130,000 square metres, of which 45,000 square metres are covered; Pioraco, which occupies a covered surface area of 15,500 square metres and employs approximately 145 people; and Castelraimondo, which has 35 employees and takes up a total area of 64,000 square metres, 15,000 square metres of which are covered.

The invention of paper, started by the Chinese and continued in the Arab world, was completed in the second half of the thirteenth century in Fabriano, a little town in the Marche region. The reason for this location, which made Fabriano the most important paper production centre of Europe, is very probably linked to the vicinity of Ancona, a port that was particularly open to trade with the Arab world. In the 1600s and 1700s, the Fabriano paper industry experienced a decline, most likely because of the shift in trade flows linked to the discovery of America. It was during the Industrial Revolution, around the same time as the French Revolution, that the industry experienced a marked upturn thanks to the entrepreneurial skills of Pietro Miliani who, in 1782, founded Cartiere Miliani. The new company boasted great efficiency and in a short space of time attained a leading position in the industry. His nephew, Giuseppe Miliani, followed the same approach, successfully transforming a family business into a major industrial complex and, above all, returning Fabriano to its privileged status on the international markets. Finally, Giovanbattista Miliani, the last great representative of the family, favoured the merger of Cartiere Miliani with the other paper mills of the area, thus creating a major industry and gaining international importance. This allowed for the acquisition of important orders from national and foreign Credit Institutes and Central Banks, thanks in the main to the company's unsurpassable skill in security paper production. On 6th June 1906, 'Società Anonima Cartiere Pietro Miliani' was founded. In 1928, a majority share was sold to Portals of London and in the 1930s, with the passing away of Giovanbattista, the Miliani family left the company management. On 15th October 1931, in order to nationalise the factory, a consortium was set up consisting of public bodies, including the Banco di Napoli, the Istituto Nazionale delle Assicurazioni (I.N.A.), the Istituto Poligrafico e Zecca dello Stato (IPZS – the Italian Mint), and Banca Nazionale del Lavoro. On 9th April 1947, at an extraordinary shareholders' meeting, the company name was changed to 'Cartiere Miliani Fabriano'. In 1972, I.N.A. became the majority shareholder, replaced by I.P.Z.S. in 1980. Then, on 21st March 2002, Cartiere Miliani Fabriano S.p.A. joined the Fedrigoni Group of Verona, which holds 99.99 per cent of the share package.

Cartiere Miliani Fabriano is one of the few firms where paper is still produced manually, using the techniques from 700 years ago. In this specific production process, the skills and knowledge of the operators are highly valued because of the strategic role played by individual competencies and capabilities. The raw material, i.e. cotton, also emanates from tradition. Nevertheless, production processes do not employ only traditional methods. In fact, they are based on the newest technological solutions. Such an attention toward technological developments demonstrates well the co-existence of old and new in the same organisation, thus revealing how innovations come from the combination of traditional and modern solutions.

The company invests large financial and human resources to appropriate the unique value of its product: paper. The firm strategically operates to strengthen the tight link that has persisted for more than eight centuries between paper production and the geographic areas of Fabriano and Pioraco, thus aiming at valorising the tradition it employs in its business. To achieve this goal, two different museums (located in Fabriano and Pioraco) have been founded with the collaboration of local institutions, which show how the art of paper production has evolved from the thirteenth century in this area.

In the Fabriano Museum it is possible to see the production process for handmade paper, and how Cartiere Miliani Fabriano has become the natural heir to this tradition. The Museum organises educational courses for children and 'Paper, Print and Book', an activity held by Lynn Sures of the Washington Pyramid Atlantic Art Centre, during which it is possible to discover the watermark, incision, printing, binding techniques and the art of the Marche region.

Ditta Bortolo Nardini

'Innovation in the name of tradition', this is the objective of Nardini, affirmed leader in the market of quality grappa made in Italy. The firm was founded in 1779, when Bortolo Nardini, already an expert in the art of distillation, left his home town Segonzano in the Cembra Valley near Trento and arrived in Bassano to open his distillery and 'grapperia' at the entrance to the famous bridge (designed by Palladio) on the Brenta river. In 1860, Bortolo Nardini, the grandson of the founder, introduced steam distillation, which became a substitute for the direct flame technique and still represents a distinctive traditional characteristic of the firm. Thanks to this innovation, the distillate retains a uniform quality that has been a hallmark of Nardini since the beginning. During the 1950s, after the Second World War, the Nardini's consumer base significantly grew as grappa conquered new markets both in Italy and abroad, affirming themselves as the only Italian traditional distillate. In 1964, a new distillery was inaugurated outside the city centre and, at the same time, a new innovative solution was introduced – vacuum distillation – which permits distillation at lower temperatures to obtain a product of superior quality. Following this, in 1981 a new transformation and bottling centre was inaugurated, and in 1991 Nardini purchased and completely refurbished the distillery of Monastier near Treviso where a substantial part of the distillation actually takes place.

Traditional competences and know-how are still employed in the distillation process, in the distilleries of both Bassano del Grappa and Treviso. In particular, in the historic distillery of Bassano, the distillation method is traditional, in a discontinuous cycle. In the cauldrons, steam flows through the fermented grape pomace and up through the distillation column, separating the heart from the 'head' and 'tail', passing onto the rectification and condensation phases. As a result of the vacuum method, the distillation takes place at lower temperatures, thereby extracting more aroma and flavour from the pomace.

The significant attention given to its own traditions is also very much in evidence in a number of initiatives Nardini has undertaken to valorise and share with customers its values and culture. For example, it is important to mention the Grapperia Nardini, which, established in 1779, has remained the same through time and claims distinction in being one of the 'Historic Places' of Italy.

A further initiative is represented by the Garage Nardini, which is a meeting of minds uniting the Nardini family's passion for contemporary art and the city of Bassano's necessity for a space to showcase the creativity of artists from all over the world, a space immortalising moments of modernity to a wide public, united in a common passion for this type of art. The use of art for strategically leveraging the firm's tradition may also be observed in the relationship between Nardini and the Guggenheim Foundation, as well as by the new research and events space (Bolle) that the Nardini family commissioned from the architect Massimiliano Fuksas to celebrate the 225th anniversary of the company.

The Kikkoman Corporation

The Kikkoman Corporation, recognised as the number one producer of soy sauce in the world, has grown from humble beginnings in the seventeenth century to become a company that provides a comprehensive range of foodstuffs, wines, dining and effective biotechnology. The company was founded in 1917 in Noda, a small city located not far from Tokyo, at the end of 2009 and has reached net sales equal to ¥412,649 million and employs 5,263 people. Noda is located on the Kanto plain, once one of the largest production centres for soybeans and wheat. It was easy to secure the necessary workforce in this region, owing to its growing population following the establishment of nearby Edo (today's Tokyo) as the capital of Japan. Besides its geographic advantages, boat transport had developed in Noda, which was blessed with two major waterways: the Tone and Edo rivers. The transportation of ingredients was thus made easier, while manufactured products could be delivered to Edo for mass consumption. Based on historical documents, 19 soy sauce brewers organised an association in Noda to ship soy sauce mainly to Edo. By the mid-nineteenth century, Noda had become the largest soy sauce producer in the Kanto region.

The history of Kikkoman and its famous soy sauce began when the Mogi and Takanashi families began soy sauce production along the Edo River in Noda. For more than three centuries, the company formed by these families has created delicious, all-natural seasonings that have won prizes worldwide, including honours at the 1873 Vienna World's Fair and the distinctive position as the official sauces for the Japanese imperial household for many years. In the mid-seventeenth century, traders had begun exports of Japanese soy sauce and exports of Kikkoman soy sauce started in the mid-nineteenth century. Nevertheless, it was only after the Second World War that Kikkoman's export business expanded significantly, thanks to the occupation forces, journalists and other foreigners who resided in Japan and became familiar with Japanese cuisine and its use of soy sauce. In 1957, Kikkoman opened its first overseas sales base in San Francisco. To meet steadily increasing demand, Kikkoman then built its first overseas production plant in the US in 1972. Following its success in the US market, Kikkoman began to establish production and sales networks throughout the world. Today, Kikkoman has three soy sauce production plants in Japan and seven abroad and sells to markets worldwide.

The Kikkoman Group now has affiliations with Del Monte, Manns Wine Co. Ltd and JFC International Inc.

Although soy sauce is still produced in Noda and remains Kikkoman's leading product, Kikkoman today represents much more than soy sauce and has expanded far beyond the banks of the Edo River. Kikkoman Soy Sauce is now produced in the United States, Europe, Singapore and Taiwan as well. Yet the name Kikkoman means more than just superior soy sauce. According to Japanese folklore, the tortoise lives for 10,000 years and thus is a symbol of longevity – what every company hopes for. Therefore, kikko (which means tortoise shell in Japanese) and man (meaning 10,000) were chosen first as the trademark for the Mogi's best soy sauce and later as the company name. In 1917, more than 250 years after their ancestors began soy sauce production, the Mogi and Takanashi families merged to form Noda Shoyu Co. Ltd. The company's name was changed to Kikkoman Shoyu Co. Ltd. in 1964, and changed once more in 1980 to Kikkoman Corporation.

The blend between tradition and innovation is much in evidence in all the phases of the soy sauce production process. Although production has progressed from handmade to fully automated, traditional techniques and methods are still employed, thus assuring the product's high quality and differentiating it from chemically-produced soy sauces. Naturally-brewed soy sauce is made using only four basic ingredients – soybeans, wheat, salt and water – whose careful selection is necessary to ensure the authentic flavour and aroma of the soy sauce. Examples of tradition in the Kikkoman soy sauce process may be found in different phases. First, the brewing process that occurs over several months and involves more than 300 components (such as flowers, fruits, whisky and coffee); as well as in the fermentation process where enzymes or other additives (such as the chemicals that break down proteins in order to speed up fermentation) are not included. Second, since its foundation, Kikkoman has been using its original Kikkoman Aspergillus, a type of fungus, to propagate koji mould. Koji mould is one of the most important elements in the manufacture of soy sauce, and plays an essential role in fermenting the ingredients: this activity is the key to its taste.

The use of tradition is not only limited to the soy sauce production process, but is also embodied in several initiatives aimed at strengthening the relationship between the Kikkoman Corporation and the Japanese tradition of soy sauce. First, it is important to mention the Kikkoman Soy Sauce Museum, where it is possible to find rich and detailed information about naturally-brewed Kikkoman soy sauce, including the history of soy sauce, details about its production process and features of soy sauce, together with recommended recipes. Second, each quarter the company publishes the *Food Forum* newsletter, which diffuses traditional Japanese food culture and history, valorising the fundamental role played by soy sauce. Third, on the Kikkoman website it is possible to find the Cookbook, where the company has collected a great number of recipes to make dishes that embrace different food cultures, employing soy sauce. Finally, to spread Japanese cuisine worldwide, the Kikkoman Corporation had two projects located in the Japan Industry Pavilion at Expo 2010 in Shanghai, called Ryoutei and Murasaki, where chefs from renowned Japanese restaurants served dishes representing the peak of Japanese dining culture, and made skilful use of fresh seasonal ingredients while they received diners with the sincere hospitality for which Japan is known.

Lavazza

With over 110 years of history behind it, Lavazza is the result of its tenacious and intuitive founder, Luigi Lavazza, and his pioneering spirit, who first invented the concept of the blend, the art of mixing coffee of different origins to obtain a harmonious, flavourful product. Today, Lavazza is the leader when it comes to quality and is the symbol of Italian espresso, thus representing the Italian identity worldwide. Despite the continuous aim of being an international player, the Lavazza Group's European roots ensure that the region acts as a key reference area. Already the market leader in Italy, Lavazza aims to be the world leader in espresso, establishing its identity as a quality Italian brand in order to guarantee excellent coffee any day, any place. The firm's statistics are surprising: a turnover of €1,121 million in 2008; over 2,300,000 imported bags of green coffee; 2,000 worldwide employees; 14 billion cups of Lavazza coffee consumed in one year worldwide; a presence in over 90 countries around the world; four production plants in Italy; 9 foreign subsidiaries and over 80 authorised distributors collaborating with Lavazza on the making and marketing of its product.

The starting point for the modern day Lavazza Company was the opening for business, on 24 March 1894, of a grocery shop. Then, Luigi Lavazza decided to focus on consolidating business and boosting the company's prestige. He realised that in order to differentiate his business, he should make a name for himself with his most prized product: coffee. He did this through the skilful use of blends, an art unknown as yet to his competitors who simply sold a single variety of coffee. The result was a boom in his business and Lavazza Company moved to Number 10 on Via San Tommaso in Turin, which today is the only café in Italy owned by Lavazza. In 1957, Luigi's children (Mario, Giuseppe and Pericle) opened a new factory on Corso Novara, thus giving rise to the coffee industry. The foundation, in 1979, of The Lavazza Centre for Coffee Studies and Research, led to the creation of the Training Centre which focused on teaching everything about Lavazza coffee. The most important innovation developed by Lavazza lies in the processing cycle, which makes it possible to handle the coffee with a vertical or gravity process. As it falls, the coffee goes through the roasting, blending and packaging phases. This is a totally innovative process in Italy and it proves to be extremely rational and efficient, guaranteeing the processing of more than 88,000 lb of coffee a day.

During the latter years, Lavazza has won the challenge of the foreign markets and has acquired Brazilian companies, such as Café Grão Nobre Ltda and Café Terra Brasil, and Indian companies, such as Barista and Fresh & Honest Café Limited. Through these acquisitions, the Group has become one of the leading players in India in the coffee shop chain sector, and has further strengthened its position as a worldwide player. Furthermore, Lavazza has opened its first Espression coffee shop in Spain in Barcelona's L'Illa Diagonal shopping centre, and a second one in America in Chicago in the John Hancock Observatory – a unique ambience with a height of more than 340 metres. Now, Giuseppe, Marco, Francesca and Antonella represent the fourth generation at Lavazza.

Lavazza uses its traditional specialisation in blending coffee to offer a complete range of products to suit all palates, consumer habits and coffee preparation methods. Its experience in the industry allows Lavazza to develop a strict system of procedures overseeing the selection of the beans, proper roasting, assembling the right blends and putting products into high-quality, long-lasting packaging. Hence, the taste of a Lavazza espresso comes from the strength of its experience, the use of the finest raw materials and advanced technological resources. Moreover, the Lavazza Company also exploited the coffee tradition of Turin and Italy. Indeed, an excellent coffee served with fragrant croissants in one of the many bars or cafés of the city is a wholesome local tradition. Turin is famous for its historic cafés, where men of academia, politicians and noblewomen have gathered since the nineteenth century. Still part of the city's culture today, a visit to one of these cafés, to sample the delights of Piedmont's patisseries in a historic setting, is an absolute must. Lavazza coffee originated right here, in a grocery store in the city centre called San Tommaso 10. Lavazza wants to spread the culture of Italian espresso throughout the world in order to be considered as the main interpreter of this tradition.

The most important initiative undertaken by the company to reinforce and valorise the Italian culture of espresso is represented by the Lavazza Training Centre. This is the world's largest coffee school with the aim of disseminating the culture of espresso by training and educating people about flavour. Being the first global network of training centres dedicated to the research and knowledge of coffee, Lavazza achieved unimaginable success by training real experts who will preserve, value and pass on the history and culture of coffee. Apart from the main centre in Turin, other training centres can be found in another 27 countries, from Europe to the Americas, from the Far East to Australia. A real network has been formed, through which Lavazza spreads the culture of coffee, as well as its origins, tasting, preparation and service. The Turin centre is responsible for setting all the good flavour standards, which are then adopted by every other centre in the network. All the Lavazza Training Centres offer courses with an innovative, creative approach, obviously based on the company's 100 years of tradition. There are specific lesson plans for each target group based on the attendees' degree of knowledge, from baristas, sales teams and distributors to new employees, consumers, journalists and opinion leaders. The effort made to capture and appropriate value through tradition is also underlined by San Tommaso 10, the first Lavazza café, where 100 years earlier Luigi Lavazza had opened his first drug store in the heart of Turin.

Nowadays, this is a place where tradition, history and the future of coffee all come together – a symbol of excellence and a showcase for the world of Lavazza. The spirit and passion of Luigi Lavazza are still alive: photos and objects allow us to discover the symbols of a century of coffee. The walls are decorated with photos of calendars and advertising campaigns giving a visual account of the history and evolution of this company founded in Turin, and which has become a worldwide name. To accompany coffee, San Tommaso 10 offers a vast daily selection of blends and coffee-based specialties, as well as croissants and sweet treats. For a number of years now, this unique place has provided a sophisticated restaurant with a menu inspired by typical recipes from classic Italian cuisine. Finally, another Lavazza initiative aimed at showing and diffusing its traditional values is represented by the exhibition 'Lavazza Design Machines: an "express" industrial reality' established in Turin, where it is possible to discover Lavazza's Italian art of self-made espresso, the energy, technical finesse and Italian-ness of the company, but more especially the design of its machines. Through the use of design the company communicates the passing of time and fleeting fashions, the advent of a new technology and the birth of a material, as well as the evolution of human beings and their working implements. The exhibition includes the book, edited by Virginio Briatore, that takes a look back at the three Lavazza systems that have generated the socio-economic phenomenon of the self-made espresso, focusing on the machines that the company has produced since 1989.

The Lego Group

The Lego Group, founded in 1932, is a privately-held company based in Billund, Denmark. The company is still owned by the Kirk Kristiansen family, who founded it in 1932. It has come a long way since its founding – from a small carpenter's workshop to a modern, global enterprise that is now, in terms of sales, the world's fifth largest manufacturer of toys. The name Lego is an abbreviation of the two Danish words 'leg godt', meaning 'play well', that represents the motto of the company. Lego is engaged in the development of children's creativity through playing and learning. Specifically, based on the world famous Lego brick, the company today provides toys, experiences and teaching materials for children in more than 130 countries. The Group has approximately 8,000 employees, with net sales that rose by 22.4 per cent from DKK9,526 million in 2008 to DKK11,661 million in 2009.

When we say Lego we think about the brick which, in both product design and technology, represents the tradition of the company. The brick is the firm's most important product and its success is acknowledged by its double nomination as 'Toy of the Century'. Various products have been developed over the years, but the foundation remains the traditional Lego brick. The brick in its present form was launched in 1958. The interlocking principle with its studs makes it unique and offers unlimited building possibilities. It is just a matter of getting the imagination going and letting a wealth of creative ideas emerge through play. A clear proof that Lego has developed all its toys around the traditional brick is that each new product remains compatible in some way with existing pieces, despite variations in the design and purpose of individual pieces over the years. Lego bricks from 1958 still interlock with those made in 2010, and Lego sets for young children are compatible with those made for teenagers. More recently, Lego has proposed two new product lines – Duplo and Technic. The former aimed at younger children was introduced in 1969, and is a range of simple blocks which measure twice the width, height and depth of standard Lego blocks. The latter is a line of interconnecting plastic rods and parts. The purpose of these series is to create novel and more advanced models with more complex moveable parts, such as machines with wheels; remaining, however, always faithful to the brick-building properties of Lego.

Many initiatives may be mentioned to explain how Lego exploits tradition to capture the created value. The Lego Ambassador Programme is a Lego community-based volunteer programme made up of adult Lego hobbyists, who share their Lego product and building passion with the worldwide Lego community and the public. The mission of the Lego Ambassadors is to work together with the Lego Group in all areas which concern the worldwide Lego community, and to represent their part of the Lego community toward the Lego Group. Each Lego Ambassador Programme cycle is of one year duration, and the current Lego Ambassador Programme cycle has 40 members from 22 different countries from all over the world. Another initiative is represented by the Lego Certified Professionals, a community-based programme made up of adult Lego hobbyists, who have turned their passion for building and creating with Lego bricks into a full-time or part-time profession. Lego Certified Professionals are not Lego employees, but they are officially recognised by the Lego Group as trusted business partners.

Another method used by the company to demonstrate that the brick can pursue its 'vision' to inspire children to explore and challenge their own creative potential is the competition it promotes. The First Lego League (FLL), established in 1998, is an international robotics team competition for children aged from 9 to 16 years. Today FLL tournaments take place in more than 40 countries worldwide with participation from over 10,000 teams. The competition has two parts: a robot game and a project. In the robot game, teams design, build, programme and test autonomous robots that must perform a series of tasks or missions. In the project, teams conduct research and create a technological or engineering solution to an aspect of the challenge and present that solution.

Another competition is the World Robot Olympiad (WRO), a counterpart to FLL, and caters to robotics fans primarily in Asia and the Middle East. However, in recent years countries such as Denmark, Norway and Sweden have also joined, and more European and Latin American countries are expected in the near future. WRO was established in 2004 by educators from the China Association of Science and Technology, Singapore Science Centre, Pohang University in Korea, and Kanagawa Institute of Technology in Japan, in cooperation with Lego Education partners in the four founding countries. In the first year, more than 4,400 teams from 13 different countries participated. Today, WRO has been launched in 20 countries throughout the region, and more than 8,000 teams took part in the 2007 challenge. Moreover, to share its values and tradition with customers, Lego founded the Lego Education Centres. They are unique open spaces where children, supported by qualified teachers, learn through hands-on and exploratory play. Lego Education supports a variety of initiatives taken by universities, government institutions and other organisations to examine trends in education, or measure the effects of using its products. For instance, the Lego Mindstorms for Early Years Education and the Design and Making Centre in Cornwall,

England, carried out a pilot study (Summer 2005) to examine how children in a nursery school environment use and respond to the Lego Mindstorms for Schools system. Finally, Lego has developed different theme parks, such as Legoland Billund in Denmark, Legoland Windsor in the UK, Legoland California in the US, and Legoland Deutschland in Germany. Each park represents an amusement experience for families and children of all ages, thus allowing Lego to link its name not only to education, but also to entertainment.

Lindt

With six production sites in Europe, two in the USA, and distribution and sales companies on four continents, Chocoladefabriken Lindt & Sprüngli AG is recognised as a leader in the market for premium-quality chocolate, offering a large selection of products in more than 100 countries around the world. During the nearly 160 years of Lindt & Sprüngli's existence, it has become known as one of the most innovative and creative companies making premium chocolate. In 2009, the total number of employees was 7,409 and the total sales amounted to CHF2.52 billion. The profound know-how, as well as the experience in the areas of product development, manufacturing and marketing accumulated by the Lindt Maîtres Chocolatiers over generations, constitutes the principal core competence of Lindt & Sprüngli and is the secret of the company's success. The Lindt brand has become the embodiment for premium chocolate of Swiss origin, uncompromising quality and long-standing tradition. Lindt & Sprüngli was established in 1845, when Confectioner David Sprüngli-Schwarz and his 29-year-old son, Rudolf Sprüngli-Ammann, dared to do something new, in a small pastry shop on Marktgasse in Zurich's old town. They decided to make chocolate.

Having grown rapidly, it was transformed into a joint-stock company in 1898 and, one year later, the company acquired the Berne-based chocolate factory including the exclusive manufacturing secrets and the famous brand from Rodolphe Lindt, who had developed 'conching' in 1879; the method of producing chocolate that was then far ahead of its time in terms of aroma and melting qualities. His fine melting chocolate, which he called 'chocolat fondant', quickly became famous and made a key contribution to the global reputation of Swiss chocolate. At the same time, the company's name was changed to Chocoladefabriken Lindt & Sprüngli AG. In 1972, the proprietary 'LSCP method' was developed and put into production. This was one of the most important developments made in the manufacturing process since the invention of the 'conche'. The process secured the quality standards and was more energy- and space-efficient than the dozens of 'long conches', the equipment previously in use.

In 1994, the transformation of the formerly successful Swiss company into an internationally successful organisation, with its headquarters in Switzerland and a modern corporate structure, was completed. Since then, the enterprise has grown steadily, transforming itself into a multinational group through progressive integration of licencees and strategic acquisitions. In particular, in 1994 Lindt acquired the popular Austrian confectionery company Hofbauer, followed by the acquisition of the traditional production company Caffarel in Italy in 1997, and by the Ghirardelli Chocolate Company in the US in 1998. The group currently has production companies in Switzerland, Germany, France, Italy, the USA and Austria, distribution and sales companies in England, Hong Kong, Spain, Poland, Canada, Australia, Sweden,

Mexico and the Czech Republic, as well as sales offices in Dubai and Ireland. In addition, Lindt also sells its products via a vast network of independent local trade partners.

The most important creations of this chocolate firm are represented by the classic and famous tablets, pralines, treats, Lindor and Excellence. A traditional product of this firm is the Dark tablet: the first recipe in the world of dark chocolate created in 1879. Today, the 100 g tablet is still part of the Lindt selection and both the recipe and the wrapper have remained unchanged. Lindor was introduced as a tablet and only in 1967 did it appear on the market in the shape of a ball as a special Christmas edition. At that time Lindor balls had a red foil wrapping with a traditional white lace pattern that made them perfectly suitable for a Christmas decoration. The new Lindor balls were so well received by consumers that in 1969 they were included in the regular Lindt product range. This was the start of a history that has now surpassed 60 years, and which has remained unchanged since it was first created and is now sold in more than 100 countries.

The now-famous Lindt melting sensation is the creation of one man, Rodolphe Lindt, who in 1879 invented something so revolutionary to the chocolate world that to this day the Maîtres Chocolatiers of Lindt are still indebted to it. The invention was the conche and its production process, i.e. the conching. A conche is a container filled with metal beads, which act as grinders. The refined and blended chocolate mass is kept in a liquid state by frictional heat. Chocolate prior to conching has an uneven and gritty texture. The conching is a process of churning and stirring the paste for long hours. The addition of cocoa butter, while the paste was warmed by internal friction, and the use of a variety of different cacao beans, in particular ratios, were paramount to the success of the chocolate. The process produces cocoa and sugar particles smaller than the tongue can detect, hence the smooth feel in the mouth. The length of the conching process determines the final smoothness and quality of the chocolate. High-quality chocolate is conched for about 72 hours, lesser grades for about four to six hours. After the process is complete, the chocolate mass is stored in tanks heated to approximately 45–50 °C (113–122 °F) until final processing. Another distinctive and traditional element which characterises Lindt production is represented by the company's Maîtres Chocolatiers, which belong to a small group of chocolate makers who still roast these premium raw materials in their own facilities. Thus, they cultivate the chocolate-making process from the bean all the way through to their final creations. Just as significantly, they have developed a very Lindt-specific and special technique, e.g. carefully roasting the beans at controlled temperatures, to bring out the rich cocoa flavour.

The strategic use of tradition as a value appropriation driver is much in evidence in the cooperation between the Chocolate Museum in Cologne and the Lindt & Sprüngli company. The Chocolate Museum invites visitors to immerse themselves in the company history, especially in the areas of chocolate production and marketing. This partnership began in March 2006 and, since then, visitors have been able to taste the finest Lindt chocolate from the chocolate fountain and to learn something about the history of the company. The museum was started in 1993, when Hans Imhoff fulfilled his dream by building 'his' chocolate museum in the Cologne Rheinauhafen for DM53 million. Walking through the three exhibition levels is like taking a tour through the 3,000-year history of chocolate – from the Aztec ambrosia, also known as the 'food of the gods', to the modern and more industrial products. Therefore, the objective of Lindt is that of asserting its position as the heir of the ancient history of chocolate.

Persol

In 1917 in Turin, in a small courtyard in Via Caboto, Giuseppe Ratti, a former photographer and owner of an optics firm called Berry, began to manufacture technically-advanced eyewear, designed to meet the needs of comfort, security and optimal viewing for aviators and sportif pilots. Thus Protector glasses were developed, protected by 14 international patents, which was adopted early on by the Italian Army and Air Force. In 1938, following on from the evolution of the Protector model, Ratti introduced the brand Persol, which means 'for the sun', thus underlining its protective function. In the late 1930s, the Meflecto system was introduced, the world's first flexible stem, which is still considered as a distinctive Persol feature. In the same period, the firm adopted as its distinctive symbol the Arrow, inspired by the sword of the ancient warriors and immediately patent-protected in several countries. In 1957, Model 649 went into production, with an entry in 1994 in the French book *Quality: scenes d'objets à l'italienne*, as being among the objects most representative of the Italian creative genius of yesterday and today. In 1995, the Persol Luxottica Group bought the brand Persol, preserving its traditional values and characteristics, as revealed by the fact that the glasses are still produced in the historic factory of Lauriano, near Turin. Nowadays, Persol has permeated into more than 40 countries and is sold in 3,500 outlets in Italy and 12,000 around the world.

The Persol production process involves a long tradition of research and quality, where several phases are still hand-made, paying great attention to quality and comfort. The relevance of tradition for Persol has been well emphasised by Fabio d'Angelantonio, Chief Marketing Officer of Luxottica, who said: 'Persol embodies the traditional values of the

Italian hand-made art, thus becoming a world reference among luxury accessories.' The traditional values of Persol's glasses are reflected in several activities sponsored by the brand, as a partner in the Torino Film Festival, where Persol was awarded Best Documentary in the Italian section Italiana.Doc, as well as its participation at the 48th Salone Internazionale del Mobile in Milan, where Persol presented 'A Work of Persol', a communication platform introduced in 2009 to promote projects from young artists characterised by accurate craftsmanship, manual and technical dexterity, and high quality standards. Further initiatives are represented by the sponsorship at Art Basel, Bolgheri Melody, Venice Film Festival and Festival della Letteratura di Mantova.

Perugina

Perugina was founded in 1907 in the historical centre of Perugia as a workshop for the manufacture of confectionery. In 1915, Perugina changed from an artisanal firm into a real industrial company, moving to the establishment of Fontivegge. Perugina opened their first store in 1919 and, within a few years, another 12 shops were established. In 1972, the company launched the 'tube' of chocolates, which has helped to characterise the tradition of Valentine's Day. Perugina is listed on the Milan Stock Exchange and is part of the Nestlé industrial group.

Building on its centuries-old tradition, the company continues to innovate and introduce new products – such as the Perugina Collection (2004), Perugina Nero (2005) and Baci in Rosso (2006) – employing techniques and methods that distinguish Perugina's identity since 1907. Tradition plays a key role in the successful strategy of Perugina, and to capture this value a number of initiatives have been promoted.

First, it is important to mention the Perugina Chocolate House, founded in Perugia in 2007, where visitors may discover the origins of chocolate, the black gold of the Maya; learn the art of chocolate making; and the history of the Perugina brand. Second, in 1997 the company set up its Perugina Historical Museum, which is visited each year by about 60,000 people and is divided into four sections:

1. from cocoa to chocolate,
2. a history of business,
3. products and sales networks,
4. communication.

Third, particular attention should be paid to the School of Chocolate, representing a concrete symbol of Perugina's knowledge and experience. The school includes 14 labour stations having all the necessary chocolate-processing tools, so that, following the instructions of expert chocolate masters, it is possible to learn the characteristics and quality of chocolate, from raw material to final product, as well as Perugina's traditional production techniques.

Pictet & Cie

Pictet & Cie is a Swiss private bank, with its headquarters in Geneva and, from its outset, has focused solely on managing the wealth of private and institutional investors. It does not engage in any form of investment banking and does not issue any commercial, mortgage or unsecured loans. At the end of 2009, Pictet & Cie employed more than 3,000 people worldwide and had approximately CHF388 billion (€292 billion) in assets under management and administration. The bank is aiming for international expansion and currently has 20 offices in some of the most important financial centres throughout the world. Pictet & Cie is a partnership owned by six managing partners who stand personally and entirely liable for the Bank's commitments. As such, the Bank enjoys total financial independence, controls the entire chain of its operations and does not need the approval of shareholders in taking decisions. The validity of its work is certified because its wealth management services are frequently mentioned in the leading rankings, and Pictet Asset Management, the institutional arm of the Group, is acknowledged the world over for its quality and performance. The Group also ranks among the leading European fund promoters and providers of fund-related solutions and products are often rewarded by Lipper or Standard & Poor's. Pictet & Cie is ranked 10th in the world by assets-under-management according to Scorpio Partnership, the UK insight and business consultancy to the global wealth industry. This most recent distinction is confirmation of Pictet & Cie's positioning as a major wealth manager at the global level. The company has consistently been ranked as one of the leading quality providers of Global Custody in industry surveys for more than 20 years. Finally, Pictet & Cie was ranked the world's top Global Custodian in the 2010 Global Custody Survey.

The first traditional element that characterises this bank is its vocation: advising its clients in the art of managing their wealth. Pictet & Cie has been managing assets for wealthy private individuals and families since 1805. Each addition to the range of services the Group offers adds to and enriches the choice of products available to its private clients. Two centuries of experience in the banking industry allows Pictet & Cie to achieve a thorough understanding of each client's specific objectives and interests. By focusing exclusively on its clients' interests, Pictet & Cie has made transparency its guiding principle. The bank executes all trades on its clients' behalf directly on open markets and therefore ensures they enjoy competitive prices. Pictet & Cie has decided to keep transaction costs transparent and to dissociate itself from the market-making system, where the spread between the buy and sell prices can be used to mask a bank's commission charges. Therefore, the clients know the exact price or rate obtained via the correspondent, as well as knowing how much the transaction costs them. In order to ensure that its clients' long-term interests are as closely aligned as possible with those of the bank, Pictet & Cie intends to continue doing business independently, in the best private banking tradition.

Indeed, the legal structure of this bank is unique. Unlike the shareholders in a limited company, whose liabilities are restricted to their share of the capital, Pictet & Cie's six partner-managers have collective and unlimited liability for the bank's commitments over their own personal wealth. They also combine ownership and management of the bank in their hands. In Switzerland, this status which entitles them to enjoy the protected and specific status of 'private banker' and to add the '& Cie' to their corporate name, is recognised only for businesses set up by individuals or as partnerships. This legal structure has been the same since its foundation.

From its two-centuries-old heritage of banking tradition, Pictet & Cie had gathered all the knowledge and experience in order to be a pioneer in the field of institutional asset management as long ago as the 1960s, and today advises benchmark institutions with the help of well-known experts in their fields. Pictet Asset Management (PAM) is the institutional investment management arm of the Pictet group, and is one of the leading independent asset managers in Europe. It encompasses all the operating subsidiaries and divisions of the Group dedicated to institutional asset management. Thanks to PAM, the bank provides specialist investment management solutions to leading financial institutions globally, such as pension funds, mutual funds, sovereign wealth funds and financial institutions. Now, the Pictet group is an asset management specialist focusing mainly on the following areas of expertise: Private Banking, Institutional Asset Management, Fund Administration and Management, Global Custody & Investor Services, Family Office and Services for Independent Asset Managers.

The bicentennial (2005) was a very special year for Pictet, a unique opportunity in which the partners decided to issue a number of historical publications: on Pictet itself, on the world's leading financial centres and on the main events to have shaped the world since the bank was originally founded. Indeed, the bank decided to publish these three historical works, each one very different from the other, but all intended to mark the occasion in a dignified and permanent manner. Two historical publications had already been issued in the past to mark the bank's 150th and 175th anniversaries. Therefore, Pictet & Cie decided to continue the historical work started by its predecessors and publish a book on the history of the bank. Particular attention was given to the past 30 years, a turning point in the bank's development. The historical publications for the bicentennial were not only about the bank itself. The bank decided to take this special occasion as an opportunity to associate the Pictet & Cie name with a treatise extending beyond the bank's own history, thereby creating a tight link between the company and financial tradition. The subject was a historical comparison of the world's international financial centres from 1805 to the present day, outlining the reasons behind their rise and, in the case of some, their fall. At the end, there is a magazine that was designed to be both educational and entertaining. There were 10 issues in all, each covering a period of 20 years, allowing the discovery of the world that its predecessors knew, lived and worked in, and ultimately conquered.

Another initiative that the company has undertaken in order to strategically exploit its traditions for competitive advantage is the establishment of the International Museum of the Calvinist Reformation in Geneva, which refers to traditional values characterising not only the bank, but also a city and, more generally, an epoch. In fact, the Reformation had a decisive impact on Geneva's political and social scene and also helped to foster the city's emergence as a business and financial centre. Calvin was among the first leaders of the Church to react to the payment of indulgences and this allowed credit to be granted upon payment of interest, but bound by high moral requirements. For these reasons, there were no usury rates and credits were cheap. Another value that Protestantism considered worthy was the protection of privacy, which gave rise to banking secrecy. The museum displays a remarkable collection of rare books and manuscripts relating to the Wars of Religion, manuscripts and paintings of austere historical figures (theologians and pastors). The museum was designed to be not only a place of remembrance but also a place that helps us to envisage the future, thanks to the evolution of the message that gave birth to the Reformation.

Shifman

Shifman has produced handmade mattress and box spring sets in all sizes for over 100 years. Shifman mattresses are among the very last that are handmade using the finest materials, thereby creating long-lasting and extremely comfortable mattresses. The logo was designed to give the idea of refinement and elegance and it shows the name of the firm and, above it, the word 'handmade'. Therefore, Shifman wants to express the close tie between the firm and the handwork. A Shifman mattress has provided rest for presidents such as Teddy Roosevelt, JF Kennedy and Bill Clinton, and for entertainers like Antonio Banderas and Eddie Murphy. Before 2008, Shifman sales were booming, growing at about 20 per cent a year for more than a decade. The company actually employs 50 workers, many of whom have been with Shifman for more than 20 years, and has a turnover of 40 million dollars.

Shifman was created in 1893 when two brothers, Abraham and Samuel Shifman, established a company built on the simple notion that not all sleep was created equal. Together they created a company committed to manufacturing superior bedding and providing the ultimate in comfort and quality to their customers. Founded in Newark, New Jersey, Shifman & Bro. Mattress Company revolutionised the mattress industry by creating a tradition of excellence in bedding. Twenty-three years later, Samuel bought the company and, with the inclusion of his three sons, Milton, Simon and Arthur, changed the company name to Shifman Bros Inc. In 1918, the company moved to its current location at One Mott Street in Newark, NJ. As the years passed, Shifman grew to become the second largest bedding producer in New Jersey. After the Second World War, although still producing the finest quality handcrafted bedding, the company was slow to adjust to changes in the industry, with consumers wanting firmer bedding, more fashionable covers and brand name recognition created by advertising. The turning point came when Michael Hammer, a bedding professional with over 22 years' experience in sales and manufacturing, first with Simmons and later with Stearns and Foster, entered the picture in 1985. After touring the factory and witnessing first-hand the tremendous quality built into a set of Shifman bedding, Hammer was anxious to purchase the operation. He pledged to continue the legacy of Abraham and Samuel Shifman and to maintain the superior standards of quality, excellence and integrity on which the company's reputation was built. Since 1985, Hammer has instituted a variety of improvements to its manufacturing procedures while maintaining the Shifman tradition. The results have shown remarkable growth year-on-year, especially with fine furniture and department stores. In 1994, Mike's youngest son, Bill, joined the team and is currently the company's president, responsible for running day-to-day operations. In 2008, the company expanded its facility by 40 per cent in order to enhance its manufacturing capabilities to meet the growing demand for its mattresses.

The key to its success remains the traditional manufacturing process, which employs techniques and design principles that have endured the test of time. More than 50 craftspeople handmake both mattress and box springs using proprietary and traditional techniques. The Shifman mattress uses only the best tempered steel innerspring and the most durable and appealing upholstery, all their work being performed under the best standards. The purpose is to deliver the best high-quality, handmade, extremely comfortable mattresses on the market. A premium Shifman mattress takes between eight and 12 hours to produce, a snail's pace compared with the average in the highly automated mattress: one every 40 minutes. It is made with 65 pounds of American cotton stuffing, lumber from Canada, twine from Italy and fabric covers from Belgium. The box springs are still made with springs, unlike lower-priced box springs, which have been renamed foundations and are boxes made of wooden slats and covered with fabric.

Shifman's collection combines the craftsmanship of yesteryear with modern-day technology and quality control.

Each mattress is two-sided and hand-tufted, no different to a Shifman mattress from 100 years ago. And the primary upholstery material is still cotton felt, produced by Shifman's 100-year-old garneting machine. But interjected with all of this old-world charm are space-age materials such as 'visco-elastic memory foam' and latex. The combination of the old- and new-world create a bedding collection with superior comfort and support. All technologies adopted by Shifman rely on traditional competencies and technical applications, as revealed by the fact that most of the company's patents are assigned to the son of Samuel Shifman, Louis Milton.

Three elements mainly characterise the traditional nature of Shifman's mattresses – Sanotuft, innerspring and tailoring. Shifman's exclusive Sanotuft technique, done entirely by hand, produces a buttonless tuft through all layers of upholstery with the innerspring locking everything securely in place, eliminating shifting or bunching. All Shifman mattresses are tufted, or bound together with nylon twine that a worker, starting with a double slip knot, threads top to bottom and back again with 24-inch steel needles, tying another knot at the end. The tufting keeps the materials inside from shifting and gives the mattress surface a dimpled look. On conventional mattresses, the top and the bottom surfaces are quilted by machine. Such dedication to the tufting, and the utmost care and attention paid to it that machine-tufting cannot possibly provide, ensures that the mattresses last long and are more resistant to wear and tear. Finally, another important value is added by the tailors of the company who carry out over 280 individual sewing operations in every queen-size mattress, Swiss-loomed embroidered borders and sewn-on handles. The use of tradition at the appropriation level is less structured than in other cases and it is mainly managed through advertising and testimonials. In this regard, the company's website provides a section dedicated to the compliments to Shifman received from both customers and partners during the years.

Steinway & Sons

Steinway & Sons was founded in 1853 by a German immigrant, Henry Engelhard Steinway, in a Manhattan loft on Varick Street. Over the following 30 years, Henry and his sons, C. F. Theodore, Charles, Henry Jr, William and Albert, developed the modern piano, patenting half of the 127 company's inventions, mainly based on emerging scientific researches, as the acoustical theories of the renowned physicist Hermann von Helmholtz. In 1866, Steinway & Sons opened the first Steinway Hall on 14th Street and, with a main auditorium of 2,000 seats, it became New York City's artistic and cultural centre, housing the New York Philharmonic until Carnegie Hall opened in 1891. By this time, the company had moved to its current location in the Astoria section of Queens, New York, and built the Steinway Village, which was a virtual village including foundries, factory, post office, parks and housing for employees. After the death of Henry, his sons Theodore and William assumed management of the operations. In 1888, due to the increasing demand for pianos from all the world, Theodore opened the Hamburg plant, which actually remains the company's only manufacturing facility outside of New York. Nowadays, Steinway & Sons crafts about 2,500 pianos a year and supports over 1,500 prominent concert artists. During the 2008/2009 concert season, 98 per cent of piano soloists chose Steinway.

Each piano maintains its original design and functionality, assured by the employment of the technological solutions developed by Henry Steinway and his sons. In addition, each piano is built following traditional rules and techniques, applying skills that were handed down from master to apprentice, generation after generation. For example, each Steinway grand piano takes nearly a year to be created. The woods employed in the rims, tops and soundboards is carefully selected and they are cured for months in the company's yard, kilns and conditioning rooms, until they stabilise at a rigidly specified moisture content.

The main initiative undertaken by Steinway & Sons to valorise its tradition is represented by the tight partnership the company has established during the years with pianists from every genre, such as Lang Lang, Diana Krall, Billy Joel, Irving Berlin, Cole Porter, Sergei Rachmaninoff and Arthur Rubinstein, who contributed to extolling the high quality and performance of its pianos around the world. Furthermore, it is important to mention the All-Steinway Schools, which, in partnership with 10 conservatoires, 94 universities and 22 other schools of distinction, provide their students and faculties with the best equipment possible for the study of music. Finally, to make customers aware of the uniqueness of its products, Steinway & Sons sales department provides customised advice and assistance that includes technical services and technician seminars.

Tong Ren Tang

Tong Ren Tang (TRT) is a famous time-honoured brand in the traditional Chinese medicine industry. The company was founded in 1669 under the Qing Dynasty and for more than 200 years was the exclusive supplier of natural medicine to China's Imperial Court. The direction of the development of TRT is: 'with modern Traditional Chinese Medicine (TCM) at the core, developing the industry of health and becoming an internationally famous modern TCM group.' TRT has been recommended as one of 16 Chinese enterprises with the strength to become a world-famous brand by the China Federation of Industrial Economics and the China Top Brand Strategy Promotion Committee. In 2006, TRT was listed in the National Intangible Cultural Heritage Directory. At present, TRT has 17 wholly-funded companies and joint ventures and 32 drugstores in 16 countries and regions. The foreign exchange earnings through export in 2008 was US$23,090,000. This has been constantly increasing, and the company is ranked first in the export of traditional Chinese medicines. TRT's products are sold to over 40 countries and regions and the company employs 2,897 people.

Tong Ren Tang sticks to the business model of a combination of medicines and treatment and has formed the overseas development pattern featuring the unique culture of TRT and integration of famous drugstores, famous medicines and famous doctors while focusing on wholesale as well as retail. The growth high of the firm is witnessed by Deloitte's 'Technology Fast 500 Asia Pacific' report. Across all 500 companies considered, TRT is in 137th position with an average revenue growth rate of 275 per cent. Today it is a US$10 billion business in China and is rapidly expanding internationally.

Tong Ren Tang began with the Yue family, whose ancestors were among the most famous 'bell practioners'. They had 'bell' in the name because they received attention from their patients by ringing the bell. Bell practitioners often used pills, pulvis, emplastrum and unguentum which they made themselves from secret recipes to cure various difficult and complicated cases of illness, yielding good results for low charges. With the development of the Yues over the generations, they accumulated a notable fortune and established their own pharmacy that people colloquially called 'the Yues' Old Pharmacy'. One day in 1669, a horizontal inscribed board was hung high in the hall by Xue Xianyang. On the background with black paint, there were three big, golden Chinese characters, Tong Ren Tang. On an auspicious day in 1702, another new store opened in the south of the road to Dashilar in the Front Gate.

The medicine of Tong Ren Tang was first adopted for imperial use in the palace in the first year of the reign of Emperor Yongzheng. The emperor authorised TRT to offer up medicinal materials to the imperial pharmacy and prepare patent medicines necessary for the royal court on behalf of the imperial pharmacy. In the feudal royal society, if a pharmacy was chosen to serve the royal court, it showed that this pharmacy had the greatest strength and the most reliable reputation and would stand high in public favour. After TRT opened for business, it began to serve the royal court from the first year of the reign of Emperor Yongzheng and went through the reigns of a further eight emperors, independently undertaking official medicines for a total of 188 years.

The turning point came with the entry into the company of Yue Pingquan, heir to the founding family. He decided to rely on a word very familiar to us today: innovation. By thoroughly sorting through old prescriptions and developing new medicines, he accumulated wealth, expanded the market and prevailed over competitors. He developed new and better kinds of medicines, and the new medicines he created greatly enriched the list of medicines of TRT, and added many precious and excellent prescriptions to the treasure-house of medicine of China. Many of these new medicines developed by Yue Pingquan are still the traditional famous brand products or important products of the company today.

The success of Tong Ren Tang greatly depends on the creation of value through tradition. A tradition made of its own recipes, patents and experience. Since its establishment, the company has been producing and using Chinese herbs guided by Chinese medicine theories. It collects and develops effective recipes and has been innovating and improving them in practice. The tradition, handed down for more than 340 years, has made this company successful, and it has become a pillar of traditional Chinese medicine. TRT produces traditional Chinese medicines using more than 5,000 substances, some of which are prepared in the form of pills and tablets or syrup, while others are made for use in decoctions. TRT wisely follows the techniques of traditional Chinese medicines whose natural origin, thanks to clinical trials and product selection over millennia, have proven to have pharmacological properties and fewer adverse effects and toxins than other compounds. Since its foundation, the company has been able to successfully produce thousands of products, divided into medicine, cosmetics and tea – all prepared from the basic herbs and ingredients strictly prescribed by the TCM. Herbal medicines are the oldest and the most famous products of this successful firm. Specifically, TRT has been engaged in the preparation of such medicines since its inception and, over time, the recipes have been handed down to their heirs in improved and updated form.

Following the Chinese tradition, three types of tea are produced by the company, all conceived as natural remedies for different types of ailments. Finally, the field of cosmetics is an area recently developed by the company, which has remained faithful to the Chinese traditional preparation of such products. Furthermore, TRT's traditional character does not depend solely on the production of medicines, but also on the employment of traditional treatments, such as herbalism, acupuncture, moxibustion and fire cupping.

An important role for determining the success of Tong Ren Tang is its appropriation strategy, aimed at developing the culture of TCM and its effectiveness – not only in China, but also abroad. A clear example is provided by the Beijing Tong Ren Tang Hospital of TCM. It is a TCM hospital combining medicine and medical treatment, founded to preserve and spread the company's and Chinese cultural values. In particular, the hospital adheres to the principle of traditional Chinese medicine supplemented by western medicine, and the principle of 'large special and small general departments'. It includes an expert outpatient service, general outpatient service, inspection centre and wards. The expert outpatient service is a window for promoting TRT's culture and it reflects the idea of serving the people and cultivating benevolence. The department appoints dozens of experts on traditional Chinese medicine with excellent medical skills and ethics in Beijing and all over the country. The combination of the academic level of famous TCM experts, its unique clinical effects, and the TCM brand have given rise to the advantage of 'integrating medicine and medical care'. Making full use of the advantage of TRT in TCM, the hospital also provides patients with a sales service. Finally, according to the disease-prevention theory of TCM, the hospital has also established a counter for food therapy and health care to provide services such as cooking soup with TCM and medicinal drinks.

Furthermore, in October 2007, Tong Ren Tang USA Corp. announced the formation of the Tong Ren Tang Wellness Corporation in partnership with Greater China Corporation, a US public company specialising in diversified growth opportunities in China and associated economies. The new company, which is owned equally by the two partners, develops spa-like wellness centres that provide treatments and products based upon China's famous TRT herbal medicines. The Tong Ren Tang Wellness Centres represent the first extension of these herbal treatments to the spa market. Wellness centres are to be located in historical Chinese communities and other retail locations, including health and exercise centres, shopping malls, airports and existing spas. A higher-end market will be developed in conjunction with luxury hotels, cruise ships and other exclusive locations. Pilot programmes are currently under development for both market segments. The growing fascination with Chinese traditional medicines and the desire for more spa-like treatments and wellness programmes represents the positive trends upon which the Tong Ren Tang Wellness Centres are to be developed.

5

Toward a new perspective of innovation

Abstract: This chapter presents a summary of the main ideas, concepts and approaches discussed throughout the book, thus tracing the pattern of an alternative innovation strategy based on the notion of tradition. In particular, it describes which type of tradition (firm, territory, or age) is employed by each one of the companies analysed, how the companies employ technology and design at both product and process level to create new value, and which initiatives they undertake to capture the created value. Finally, a set of guidelines that firms may follow to implement a tradition-based strategy are provided, ranging from identifying which traditions to exploit, to realising initiatives for sharing with customers the uniqueness of that tradition's values and meaning.

Key words: types of tradition, tradition-based appropriation initiatives, tradition-based strategy guidelines.

Nowadays, several factors may contribute to explaining the necessity of rethinking innovation and moving toward new models and paradigms. First, modern society is currently engaged in a complex crisis, commonly known by the term 'triple crisis' (Tarp and Addison, 2010). This includes different dimensions, such as: the financial one which started in 2007 and has triggered a global economic downturn; the environmental question which is mainly concerned with problems relating to climate change, municipal waste generation and food scarcity, which are jointly affecting poverty, inequality and human development. Second, existing knowledge, competencies and skills are often under-exploited, leading to the abandonment of still-useful capabilities and to a crazy search for novelty. Obviously, such behaviour is expensive, both in economic and environmental terms, and it cannot be sustained in the long term by companies or countries as a whole. Furthermore, the above-mentioned search for novelty is often led by technological development. In fact, technology has always been recognised as a fundamental driver, favouring civilisation and allowing firms to improve existing products and processes, and to develop new ones.

By exploiting new technological solutions, firms may create new values for markets and customers and, thus, compete globally. However, technology does not always permit the building-up of distinctive resources and the sustaining of a competitive advantage, since it is apt to suffer from an inevitable diffusion and flattening process (see the convergence toward dominant designs) that limits its rarity and inimitability. Therefore, firms are forced to search constantly for distinctiveness, enhancing the technological value of products with something that contributes to making them unique. Third, the economic landscape has a profound impact on consumers' choices. New appetites have emerged and markets have sprung up to serve them. Consumers can afford to be curious about gadgets and technology, shell out for enriching (or just fun) experiences and indulge themselves with premium products. They can afford to pay extra for socially-conscious consumption, while winking at purchases that might not be squeaky clean but that they feel they deserve. The current recession has not so much put an end to this party as sobered it up, propelling some consumer trends forward while slowing, halting or even reversing others in ways that will affect their trajectory during and after the recovery. Specifically, consumers are showing an increasing demand for simplicity and paying more attention toward the ethical behaviour of companies, as well as revealing a desire for a more wholesome and less wasteful life (Flatters and Willmott, 2009).

Fourth, the loss of identity is becoming a central question that needs to be addressed not only at the societal level, but also at the company level. Individuals are rediscovering the importance of belonging to a specific national or regional culture. Identity includes a sense of continuity, a sense of uniqueness from others and a sense of affiliation, which allows individuals to move from an undifferentiated perception of themselves to a differentiated one. This tendency is also extremely evident in the current re-conceptualisation of the globalisation process, which implies the necessity for promoting and sustaining local differences and peculiarities (Stone, 2001).

In order to provide a possible answer to the above-mentioned issues, this book presents a new perspective on innovation patterns, focusing on the strategic role that may be played by tradition as a source upon which to build novelty and shape competition. Following the recombinant view of innovation, which sees knowledge creation as a recombination associated with the fabrication of a new product, process or market by combining existing areas of knowledge (Kogut and Zander, 1992), this research intends to figure out the importance of combining and exploiting old and established competencies and capabilities, as well as the impact that this may exert on innovation. In line with this theoretical perspective of innovation, the book puts forward the notion of tradition – defined as the whole set of competencies, knowledge, values and culture that characterises a specific firm – as well as a territory and an age. Specifically, by exploiting tradition, a firm selects distinctive old components, belonging to its history or to a specific territory and age, transfers them over time and incorporates these old elements into its current production processes and business practices, thus creating new value for customers. In other words, by combining hard and soft features, tradition allows the firm to distinguish itself from its competitors in terms of both meaning and content, to promote uniqueness and, consequently, to achieve a competitive position.

The strategic role played by tradition relies on the fact that it may enhance a firm's capability to both create new value for customers and to appropriate this value. As discussed previously, by renewing the past, firms may follow the new trends of customers, which are becoming especially important in the current scenario, through enrichment of the technological content of products and processes with meanings and values that help to satisfy emotional and socio-cultural needs. Moreover, tradition is sticky and embedded into specific and distinct contexts, which limits others from imitation and allows the firm to obtain abnormal economic profits and a competitive advantage. A firm's tradition relies on the notion proposed by Nelson and Winter (1982) of organisational routines that represent the store of the organisation's specific operational knowledge. Routines describe the 'way of doing things', distinguish each organisation from the others, and are the means by which a firm can preserve its organisation's knowledge base. Therefore, firms depend upon their own routines and may be expected to behave in the future according to their past routines. Such a distinction still remains when dealing with tradition at geographical level. In the marketing literature, this issue is generally known as the provenance paradox (Desphandé, 2010).

A product's territory of origin establishes its authenticity and consumers tend to associate certain geographies with the best products: French wine, Italian sports cars, Swiss watches. Competing products from other countries are perceived as less authentic, and even when their quality is on a par with that of established players, the developing-market firms are unable to command a fair price. The lower price, in turn, reinforces the idea that the offering is not as good and that the region does not make premium products. Thereby, this spatial dimension of tradition contributes to an explanation of why 'made in Brazil' implies high-quality coffee but not high-quality aircraft; 'made in China' implies high-quality silk but not high-quality financial services; Chocolates El Rey suggest that although Venezuela produces the world's best cacao, it is not considered a legitimate source of great chocolate. Tradition creates a protected geographical status, which may assure a kind of monopolistic power in certain markets, as well as significant difficulties for competitors coming from other countries to be competitive. For example, when Corona beer entered developed markets, it was disparaged by many in Europe as more fattening than European beer. Some derided it as 'Mexican lemonade'. In 1987, Corona filed a lawsuit against a US Heineken distributor for spreading rumours that Mexican brewery employees regularly urinated into the product during the brewing process. Grupo Modelo, Corona's maker, worked painstakingly to counter such claims and educate consumers, while also aggressively branding the product as a lifestyle beer and downplaying its Mexican origin. Tradition may be strategically exploited also when it refers to an age. Recently, a study conducted by McKiney & Co. (Ferrari and Goethals, 2010) emphasised the importance of the Italian Renaissance, stressing how it can still teach rules for effectively stimulating innovation. In particular, the authors identified a parallel between the Renaissance and principles that have proven their worth in R&D organisations, such as a collision between different experts, providing loose guidelines and establishing stretch goals. The lesson they propose is based on the fact that the Renaissance creativity was built on professional rivalries, such as the ones between Leonardo, Michelangelo, Raphael and Titian. In a similar manner, by sustaining and using rivalry, R&D organisations may promote the development of groundbreaking innovations.

The book proposes an approach for creating new value through tradition based on the leveraging of two main drivers: technology and design. Furthermore, both technology and design may be applied to develop products or processes. The use of traditional technology at product level refers to the functionality given to products by old technological solutions and techniques that may be able to better address specific customer needs. A traditional design is related to the semantic dimension of products, especially to the meaning and values that it transfers to customers. At the process level, using traditional technologies and design generally implies the application of artisanal techniques, as well as ways of organising production, that contribute to distinctively characterising processes by reducing their standardisation and increasing the differences and, thus, the value that each customer may obtain. Nevertheless, exploiting traditional technology and design at both product and process level is not enough to make such a use profitable and to ensure a competitive advantage. In fact, a firm needs to capture the created value, limiting others from imitation and enhancing margins and profits from the final products. To address this goal, the firm has to establish a tight relationship between products/processes and its identity, reflecting tradition in its organisational routines. In other words, tradition must assume a central role within the firm, affecting its way of producing goods and services, as well as its mission and vision. In so doing, a firm may be able to limit others from imitating and sharing with customers the traditional values and meanings that contribute to making it different from competitors.

Firms that use tradition for both creating and appropriating value may be defined as tradition-based. In this book, we focus our attention on this specific category of companies, highlighting which kind of tradition they employ (firm, territory, age), which drivers they use to embed tradition into products and processes (technology, design) and which initiatives are undertaken to increase appropriability. Twenty companies have been discussed, all tradition-based firms operating as leaders in the international markets. Table 5.1 shows the companies discussed, the relevant industries and their locations.

The first consideration that emerges from the analysis is that, in all the different industries where the companies operate, tacit knowledge seems to exert a fundamental role. In these sectors, competencies and knowledge are still understood and they are strongly embedded in the tradition of behaving and operating according to certain rules. The competitive advantage largely depends on human skills, which affect a firm's capability to innovate and to be successful. The development of innovative solutions is led by the known capabilities of individuals, which could be what solely inspires innovation and effectively utilises it. Thereby, companies exploit tradition to hand down and incorporate into their organisational routines these known elements, building their competitiveness upon them.

Cases also reveal how different types of tradition are

Table 5.1 The companies, industries and countries of origin

	Company	Industry	Country
4.1	Aboca	Health	Italy
4.2	Amarelli	Food	Italy
4.3	Apreamare	Naval	Italy
4.4	Barilla Group	Food	Italy
4.5	Breguet	Watches	Switzerland
4.6	Brunello Cucinelli Group	Clothing	Italy
4.7	Caffè Vergnano	Coffee	Italy
4.8	Cartier	Jewellery	France
4.9	Cartiere Miliani Fabriano	Paper	Italy
4.10	Ditta Bortolo Nardini	Distillery	Italy
4.11	The Kikkoman Corporation	Food	Japan
4.12	Lavazza	Coffee	Italy
4.13	Lego Group	Toys	Denmark
4.14	Lindt	Chocolate	Switzerland
4.15	Persol	Optical	Italy
4.16	Perugina	Chocolate	Italy
4.17	Pictet & Cie	Financial services	Switzerland
4.18	Shifman	Bedding	USA
4.19	Steinway & Sons	Musical instruments	USA
4.20	Tong Ren Tang	Pharmaceutical	China

employed by firms. Companies have been shown to leverage their own tradition as well as traditional elements differentiating a territory and, under some circumstances, also a specific age. In Table 5.2, companies are distinguished according to the type of tradition they mainly exploit.

As previously discussed, tradition is embedded in products and processes by means of technology and design. Regardless of the type of tradition (firm, territory, age), our analysis clearly shows that companies use technological and design elements to traditionally characterise their business,

Table 5.2 **The companies and types of tradition**

	Company	Tradition
4.1	Aboca	Territory
4.2	Amarelli	Firm, Territory
4.3	Apreamare	Territory
4.4	Barilla Group	Territory
4.5	Breguet	Firm, Territory
4.6	Brunello Cucinelli Group	Territory, Age
4.7	Caffè Vergnano	Firm
4.8	Cartier	Firm
4.9	Cartiere Miliani Fabriano	Firm, Territory
4.10	Ditta Bortolo Nardini	Firm
4.11	The Kikkoman Corporation	Firm, Territory
4.12	Lavazza	Firm, Territory
4.13	Lego Group	Firm
4.14	Lindt	Firm, Territory
4.15	Persol	Firm
4.16	Perugina	Firm
4.17	Pictet & Cie	Firm, Territory
4.18	Shifman	Firm
4.19	Steinway & Sons	Firm
4.20	Tong Ren Tang	Firm, Territory

differentiating themselves from competitors and creating new value for customers. Following this approach, the companies are able to provide products with traditional functions and meanings, as well as rediscovering and valorising the artisanal nature of their production processes. In Figure 5.1, the different strategic options firms adopt to create value by tradition are given.

As for the creation of new value, tradition is strategically employed as an appropriation strategy for capturing this. In

Figure 5.1 **Tradition-based value creation strategies**

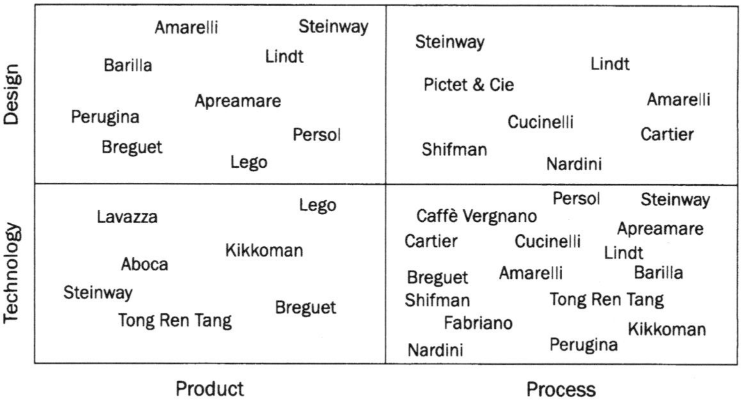

this regard, several initiatives have been undertaken by the companies in order to establish a tight relationship between their business and the tradition they employ. Historical archives and museums are the most widely adopted solutions. However, in some cases (see for example Barilla, Brunello Cucinelli and Tong Ren Tang) more sophisticated approaches are chosen, mainly devoted to capturing the benefits arising from the exploitation of a tradition that is not typical of the firm, but rather of a territory or of a specific age. In Table 5.3, the different appropriation initiatives mainly utilised by the firms are given.

In conclusion, this book sheds new light on strategic approaches firms may adopt to face competition, i.e. focusing on the recovery and employment of tradition as a distinctive resource. In particular, we argue that by employing values and competencies shaping a specific tradition, both internal and external to the firm, as characteristic of a specific territory or age, a firm may be able to distinguish products and processes, thereby creating new value for customers and appropriating it. Doing it properly, however, is not cost-free

Table 5.3 The companies and their tradition-based appropriation initiatives

	Company	Appropriation initiatives
4.1	Aboca	Herb Museum
4.2	Amarelli	Giorgio Amarelli Museum
4.3	Apreamare	Advertising
4.4	Barilla Group	Academia Barilla
4.5	Breguet	Breguet Museum
4.6	Brunello Cucinelli Group	Requalification of Solomeo Village; Forum of Arts Foundation
4.7	Caffè Vergnano	Caffè Vergnano 1882; Accademia Vergnano
4.8	Cartier	Fondation Cartier pour l'Art Contemporain; Cartier Collection; Cartier Archives
4.9	Cartiere Miliani Fabriano	Fabriano Museum; Pioraco Museum
4.10	Ditta Bortolo Nardini	Grapperia; Bolle architecture
4.11	The Kikkoman Corporation	Soy Sauce Museum
4.12	Lavazza	San Tommaso 10; Lavazza Training Centre
4.13	Lego Group	Lego Ambassador Program; Lego Certified Professionals; First Lego League; World Robot Olympiad; Lego Education Centers; Lego Thematic Parks
4.14	Lindt	Chocolate Museum
4.15	Persol	Art-Based Initiatives
4.16	Perugina	Perugina Historical Museum; Perugina Chocolate House
4.17	Pictet & Cie	Historical Publications; International Museum of the Calvinist Reformation
4.18	Shifman	Advertising; Customers' and Partners' Testimonials
4.19	Steinway & Sons	Partnership with Artists; All-Steinway Schools; Customised Assistance
4.20	Tong Ren Tang	Beijing Tong Ren Tang Hospital of TCM; Tong Ren Tang Wellness Corporation

and requires suitable skills and motivations. In the following, we suggest a set of guidelines that firms may follow in order to implement a tradition-based strategy:

- identify which tradition to exploit (firm, territory, age-specific);
- understand and recognise which distinctive elements (both tangible and intangible) of tradition may contribute to creating value for customers;
- invest in resources to embed such traditional elements into business;
- transfer tradition in products and/or processes by means of technology and/or design;
- create a tight link between tradition and company identity, reinforcing the brand and promoting communication and marketing activities;
- realise initiatives to share with customers the uniqueness of tradition's values and meaning (museums, historical archives and more complex projects for the transmission and protection of knowledge, such as the Academia Barilla), thus appropriating tradition and the value it generates.

References

Adner, R. and Snow, C.D. (2010) Bold retreat. A new strategy for old technologies. *Harvard Business Review*, March: 76–81.

Argote, L. (1999) *Organizational learning: Creating, retaining and transferring knowledge*. Boston, MA: Kluwer.

Argote, L., Beckman, S. and Epple, D. (1990) The persistence and transfer of learning in industrial settings. *Management Science*, 36: 14–154.

Arthur, W.B. (2009) *The Nature of Technology. What it is and how it evolves*. New York: Free Press.

Barney, J. (1991) Firm resources and sustained competitive advantage. *Journal of Management*, 17: 99–120.

Benkard, C.L. (2000) Learning and forgetting: The dynamics of aircraft production. *American Economic Review*, 90: 1034–54.

Bimber, B. (1996) 'Three faces of technological determinism', in M.R. Smith and L. Marx (eds). *Does Technology Drive History?* Cambridge, MA: MIT Press.

Braudel, F. (1985) *La Mediterranée*. Paris: Flammarion.

Brown, S. (1995) *Postmodern Marketing*. London: Routledge.

Brown, S. (1997) *Postmodern Marketing Two: Telling Tales*. London: ITBP.

Brown, S. (1999) Retro-marketing: Yesterday's tomorrows, today! *Marketing Intelligence & Planning*, 17: 363–76.

Csikszentmihalyi, M. and Rochberg-Halton, E. (1981) *The Meaning of Things: Domestic symbols and the self*. New York: Cambridge University Press.

de Treville, S. and Trigeorgis, L. (2010) It may be cheaper to manufacture at home. *Harvard Business Review*, October: 84–7.

Desphandé, R. (2010) Why you aren't buying Venezuelan chocolate. *Harvard Business Review*, December: 25–7.

Eliot, T.S. (1919) *Tradition and the individual talent.* First published, in two parts, in *The Egoist* magazine.

FAO (2010) *The State of Food Insecurity in the World 2010*, *www.fao.org/publications/sofi/en/2011*.

Ferrari, B.T. and Goethals, J. (2010) Using rivalry to spur innovation. *McKinsey Quarterly*, May: 1–12.

Flatters, P. and Willmott, M. (2009) Understanding the post-recession consumer. *Harvard Business Review*, July–August: 106–12.

Fleming, L. (2001) Recombinant uncertainty in technological search. *Management Science*, 47: 117–32.

Fleming, L. and Sorenson, O. (2004) Science as a map in technological search. *Strategic Management Journal*, 25, 909–28.

Freddi, D. (2009) The integration of old and new technological paradigms in low- and medium-tech sectors: the case of mechatronics. *Research Policy*, 38: 548–58.

Friedman, J. (1992) The past in the future: history and the politics of identity. *American Anthropologist*, 94: 837–59.

Gambardella, A. (1995) *Science and Innovation: The US pharmaceutical industry in the 1980s*. Cambridge: Cambridge University Press.

Gilfillan, S.C. (1935) *Inventing the Ship*. Chicago, IL: Follett Publishing.

Guerrero, L. (2009) Consumer-driven definition of traditional food products and innovation in traditional foods. A qualitative cross-cultural study. *Appetite*, 52: 345–54.

Guptara, P. (2010) Toward creating the right kind of globalization: an analysis – with proposals. *Organisational Transformation and Social Change*, B: 89–103.

Gusfield, J.R. (1967) Tradition and modernity: Misplaced polarities in the study of social change. *American Journal of Sociology*, January: 351–62.

Habermas, J. (1970) *Toward a Rational Society*. Boston, MA: Beacon Press.

Hall, J.M. and Johnson, M.E. (2009) When should a process be art, not science? *Harvard Business Review*, March: 58–65.

Haynes, P. (1998) Nostalgia. *Forbes*, 26 January: 47.

Heilbroner, R. (1996) 'Technological determinism revisited', in M.R. Smith and L. Marx (eds), *Does Technology Drive History?* Cambridge, MA: MIT Press.

Hutton, R. (1999) Time machines. *The Sunday Times*, Sport, 7 March: 23.

IIT Research Institute (1968) *Technology in Retrospect and Critical Events in Science*. Washington, DC: National Science Foundation.

Inglehart, R. and Baker, W.E. (2000) Modernization, cultural change and the persistence of traditional values. *American Sociological Review*, February: 19–51.

Kanter, R.M. (1995) Thriving locally in the global economy. *Harvard Business Review*, August: 119–27.

Katila, R. (2002) New product search over time: Past ideas in their prime? *Academy of Management Journal*, 45: 995–1010.

Katila, R. and Ahuja, G. (2002) Something old, something new: A longitudinal study of search behavior and new product introduction. *Academy of Management Journal*, 45, 1183–94.

Katila, R. and Chen, E.L. (2008) Effects of search timing on innovation: The value of not being in sync with rivals. *Administrative Science Quarterly*, 53: 593–625.

Kogut, B. and Zander, U. (1992) Knowledge of the firm, combinative capabilities and the replication of technology. *Organization Science*, 3: 383–97.

Lee, H.L. (2010) Don't tweak your supply chain – rethink it end to end. *Harvard Business Review*, October: 62–9.

March, J. (1991) Exploration and exploitation in organizational learning. *Organization Science*, 2: 71–87.

Margolin, V. and Buchanan, R. (1995) *The Idea of Design*. Cambridge, MA: MIT Press.

Marx, L. and Smith, M.R. (1996) 'Introduction', in M.R. Smith and L. Marx (eds), *Does Technology Drive History?* Cambridge, MA: MIT Press.

Mathews, G. (2000) *Global Culture/Individual Identity. searching for home in the cultural supermarket*. London: Routledge.

Mazzucato, M. and Dosi, G. (2006) 'Introduction', in M. Mazzucato and G. Dosi (eds), *Knowledge Accumulation and Industry Evolution. The case of Pharma-Biotech*. Cambridge: Cambridge University Press.

Ministero Agricoltura (1999) *Decreto Legislativo* 30 aprile 1998 n. 173. *Decreto Ministero Agricoltura* 8 settembre 1999 n. 350.

Mizik, N. and Jacobson, R. (2003) Trading off between value creation and value appropriation: The financial implications of shifts in strategic emphasis. *Journal of Marketing*, 67: 63–76.

Naughton, K. and Vlasic, B. (1998) The Nostalgia Boom. *Business Week*, March 23: 58–64.

Nelson, R. and Winter, S. (1982) *An Evolutionary Theory of Economic Change*. Cambridge, MA: Harvard University Press.

Nelson, R.R. (1991) Why do firms differ and how does it matter? *Strategic Management Journal*, 12: 61–74.

Nerkar, A. (2003) Old is gold? The value of temporal exploration in the creation of new knowledge. *Management Science*, 49: 211–29.

OECD (1997) *Sustainable Consumption and Production: Clarifying the concepts*. Paris: OECD.

Ogburn, W.F. (1922) *Social Change with Respect to Culture and Original Nature*. New York: B.W. Huebsch, Inc.

Pisano, G.P. (2006) Profiting from innovation and the intellectual property revolution. *Research Policy*, 35: 1122–30.

Pisano, G.P. and Teece, D. (1994) Dynamic capabilities: An introduction. *Industrial and Corporate Change*, 3: 537–56.

Rosenberg, N. and Birdzell, L.E. (1986) *How the West Grew Rich. The economic transformation of the industrial world*. New York: Basic Books.

Rosenkopf, L. and Almeida, P. (2003) Overcoming local search through alliances and mobility. *Management Science*, 49: 751–66.

Schumpeter, J. (1934) *Theory of Economic Development*. Cambridge, MA: Harvard University Press.

Schumpeter, J. (1939) *Business Cycle*. New York: McGraw-Hill Book Company, Inc.

Schumpeter, J. (1942) *Capitalism, Socialism and Democracy*. New York: Harper & Row.

Sennet, R. (2008) *The Craftsman*. New Haven, CT and London: Yale University Press.

Settis, S. (2004) *Futuro del 'Classico'*. Torino: Einaudi.

Shils, E. (1971) Tradition. *Comparative Studies in Society and History*, 13: 122–59.

Smith, K., Grimm, G., Gannon, M. and Ghen, M.-J. (1991) Organizational information processing, competitive responses and performance in the US domestic airline industry. *Academy of Management Journal*, 34: 60–85.

Steinmetz, A. (2008) Competition, innovation and the effect of knowledge accumulation, *BGPE Discussion Paper*, n. 53.

Stone, D. (2001) Think global, act local, or think local, act global? Knowledge production in the global agora. *Reshaping Globalization: Multilateral Dialogues and New Policy Initiatives*, Central European University Conference, Budapest, Ungheria, 17 October.

Sutton, J.R. (1984) Organizational autonomy and professional norms in science: A case study of the Lawrence Livermore laboratory. *Social Studies of Science*, 14: 197–224.

Tarp, F. and Addison, T. (2010) *The Triple Crisis–Finance, Food and Climate Change*, 2011 *http://ourworld.unu.edu/en/the-triple-crisis-finance-food-and-climate-change/.*

Taylor, J., Wacker, W. and Means, H. (1998) *The 500-Year Delta. What happens after what comes next.* New York: HarperCollins.

Teece, D.J. (1986) Profiting from technological innovation: Implications for integration, collaboration, licensing and public policy. *Research Policy*, 15: 285–305.

Teece, D.J., Pisano, G.P. and Shuen, A. (1997) Dynamic capabilities and strategic management. *Strategic Management Journal*, 18: 509–33.

The Economist (1998) *Return of the Beetle.* 10 January: 80.

UNEP (2011) *Green Economy Report: A preview*, 2011 *www.unep.org/greeneconomy/ResearchProducts/tabid/1394/Default.aspx.*

Weber, M. (1958) *The Protestant Ethic and the Spirit of Capitalism.* Translated by T. Parsons. Reprint, New York: Charles Scribner's Sons.

Wernerfelt, B. (1984) A resource-based view of the firm. *Strategic Management Journal*, 5: 272–80.

Winter, S.G. (2006) The logic of appropriability: From Schumpeter to Arrow to Teece. *Research Policy*, 35: 1100–6.

Index

Made in the USA
Coppell, TX
18 October 2026

93262780R00085